Geometrical Dimensioning and
Tolerancing for Design,
Manufacturing and Inspection

Geometrical Dimensioning and Tolerancing for Design, Manufacturing and Inspection

A Handbook for Geometrical Product Specification using ISO and ASME Standards

Second edition

Georg Henzold

ELSEVIER

AMSTERDAM • BOSTON • HEIDELBERG • LONDON • NEW YORK • OXFORD
• PARIS • SAN DIEGO • SAN FRANCISCO • SINGAPORE • SYDNEY • TOKYO

Butterworth-Heinemann is an imprint of Elsevier

Butterworth-Heinemann is an imprint of Elsevier
Linacre House, Jordan Hill, Oxford OX2 8DP, UK
30 Corporate Drive, Suite 400, Burlington, MA 01803, USA

First edition published by John Wiley and Sons Ltd 1995
Second edition published by Elsevier Ltd 2006
Reprinted 2009

British Library Cataloguing in Publication Data
A catalogue record for this book is available from the British Library

Library of Congress Cataloging-in-Publication Data
A catalog record for this book is available from the Library of Congress

ISBN: 978-0-7506-67388

For information on all Butterworth-Heinemann publications
visit our website at www.elsevierdirect.com

Transferred to Digital Printing 2010

Contents

Preface

Technical developments, leading to higher performance, higher efficiency, lower pollutions and greater reliability, and the constraints imposed by economy and the need for rationalization, greater cooperation with licensees and subcontractors, make it necessary to have completely toleranced drawings suitable for the function, for manufacturing and inspection.

"Completely toleranced" means that the geometries (form, size, orientation and location) of geometrical elements of the workpiece are completely defined and toleranced. Nothing shall be left to the individual judgement of the manufacturer or inspector. Only with this approach can proper functioning be verified and the possibilities available for economization and rationalization in manufacturing and inspection be fully utilized.

This book presents the state of the art regarding geometrical dimensioning and tolerancing. It describes the international standardization in this field, which is laid down in ISO Standards. It indicates the deviations between the American Standard ASME Y14.5M, the former East European Standards and the ISO Standards. It describes the additional specifications laid down in the German Standards (DIN Standards). Possible further developments in the field of geometrical tolerancing are also pointed out.

What has to be regarded for manufacturing within the geometrical dimensioning and tolerances is explained.

Principles for the inspection of geometrical deviations are given, together with a basis for tolerancing suitable for inspection.

Examples for tolerancing appropriate to various functional requirements, a guide for geometrical tolerancing, are described.

This book may serve as an introduction to geometrical dimensioning and tolerancing for students, and may also help practitioners in the fields of design, manufacturing and inspection.

The author has used his experiences gathered during the elaboration of the ISO Standards and during his lectures and discussions on geometrical dimensioning and tolerancing in industry and education. However, proposals for improvements are appreciated and should be directed to the publisher.

The book represents the author's understanding of the various standards and is written for educational purposes. In cases of dispute, the original standards have to be considered. The functional cases of geometrical requirements vary over a wide range. Therefore, before application in practice, it should be checked whether the presentation in the book fits the particular purpose in the practical case.

Standardization is a dynamic and continuous process. The standards follow the development in engineering. Future changes in standardization may make it necessary to update some content of this book.

Georg Henzold was manager of the department for standardization of a manufacturer of power plant machinery. He is chairman of the committee dealing with the standardization in the field of geometrical dimensioning and tolerancing in the German Standardization Institute (DIN), and he was the former chairman of the pertinent European Committee for Standardization (CEN). He is a long-time delegate in the pertinent committees of the International Standardization Organization (ISO).

Georg Henzold

Acknowledgements

To Eric Green, C. Eng, MI. Mech. E., SES, P. Eng., Ottawa, Canada, who volunteered to take over the great burden of editing the text into the English language.

To my wife Ingeborg who encouraged me and gave me her support and help.

Notation

A reading, measured value
B zone
C mean size
H height
I actual size
L length
M_a maximum material size (MMS)
M_i least material size (LMS)
N number
P mating size or content or statistical probability or pitch
R calculated value from readings
T size tolerance
V_a maximum material virtual size (MMVS)
V_i least material virtual size (LMVS)
Z centre

a deviation or variation
b width
c distance or safety factor with statistical tolerancing
d diameter
e local deviation
h height
k correction factor
l length
n number
p coordinate
q coordinate
r radius
s clearance
t geometrical tolerance
u measuring uncertainty
x coordinate
y coordinate
z coordinate

Δ difference
α angle
β angle

γ angle
δ deviation corresponding to ISO 1101
λ cut-off, wavelength

Subscripts

a   coaxiality or arithmetical
b   profile of a line
c   position or section
d   orientation
e   flatness or external
f   form
g   straightness
h   profile of a surface
i   internal
k   crossing
l run-out or left
m   measured value or centre
n   perpendicularity
o   location
ö   local
p   parallelism or point
q   longitudinal section profile
r   roundness or right
s   symmetry or statistical
t   total run-out
th   theoretically exact
u   orientation or location or below
v   twist
w   angularity
x x direction
y y direction
z z direction
z   cylindricity

Abbreviations

AVG	average
BASIC	theoretical exact dimension
CMM	coordinate measuring machine
DIA	diameter
D&T	dimensioning and tolerancing
FIM	full indicator movement
FIR	full indicator reading
FRTZF	feature-relating tolerance zone framework
GD&T	geometrical dimensioning and tolerancing
GPS	geometrical product specification
LD	least (minimum) diameter
LMC	least material condition
LMR	least material requirement
LMS	least material size
LMVC	least material virtual condition
LMVS	least material virtual size
LSC	least-squares circle
MCC	minimum circumscribed circle
MD	major diameter
MIC	maximum inscribed circle
MMC	maximum material condition
MMR	maximum material requirement
MMS	maximum material size
MMVC	maximum material virtual condition
MMVS	maximum material virtual size
MPE	maximum permissible error (of a measuring device)
MZC	minimum zone circle
PD	pitch diameter
PLTZF	pattern-locating tolerance zone framework
PUMA	procedure for uncertainty management
RFS	regardless of feature size
SEP REQT	separate requirement
SIM REQT	simultaneous requirement
TED	theoretically exact dimension
TIR	total indicator reading
TP	true position, theoretically exact position or location
VD&T	vectorial dimensioning and tolerancing

ISO Text Equivalents

A new ISO Standard (ISO 14 995-1) on text equivalents is in preparation. The standard will probably give the following text equivalents (to be used in text, not in drawings):

Symbol	Name	Abbreviation
⌒	Line profile	PFL
⏤	Straightness	STR
○	Roundness	RON
⌓	Surface profile	PFS
⏥	Flatness	FLT
⌭	Cylindricity	CYL
∠	Angularity	ANG
//	Parallelism	PAR
⊥	Perpendicularity	PER
⌖	Position	POS
◎	Coaxiality	CAX
⌰	Symmetry	SYM
↗	Circular run-out	CRO
↗↗	Total run-out	TRO

New ISO Terminology

New ISO standards on terminology are in preparation. These standards will probably define the following:

feature of size: geometric shape defined by a linear or angular dimension which is a size (cylinder, sphere, two parallel opposite surfaces, cone, wedge)

integral feature: surface or line on a feature

derived feature: centre point, median line or median surface derived from one or more integral features

nominal integral feature: theoretically exact integral feature as defined, for example, by a technical drawing

nominal derived feature: centre point, median straight line or median plane derived from one or more nominal integral features (former: **nominal centre point, nominal axis, nominal median plane**)

real surface of a workpiece: set of features which physically exist and separate the entire workpiece from the surrounding medium

real (integral) feature: integral feature part of a real surface of a workpiece limited by the adjacent real (integral) features

extracted integral feature: approximated representation of the real (integral) feature, obtained by extracting a finite number of points from the real (integral) feature; this extraction is performed in accordance with specified conventions (several such representations may exist for each (integral) feature) (former: **actual surface, actual surface line**)

extracted derived feature: centre point, median line or median surface derived from one or more extracted integral features (extracted axis, extracted median surface) (former: **actual centre point, actual axis, actual median surface**)

associated integral feature: integral feature of perfect form associated with the extracted integral feature in accordance with specified conventions (former: **substitute element, least-squares element, minimum zone element, contacting element, maximum inscribed element, minimum circumscribed element**)

associated derived feature: centre point, median straight line or median plane derived from one or more associated integral features.

*As the new ISO standards are not yet finalized the old (**bold**) terms are still used throughout the text of this book.*

For more detailed information see clause 23 ISO Geometrical Product Specification (GPS), new approach.

ISO drawing rules

The ISO drawing rules are laid down in ISO 128 and ISO 129.

1

Properties of the Surface

The suitability of a workpiece for its purpose depends on the inner properties (material properties, inner discontinuities, e.g. shrink holes, inner imperfections, e.g. segregations) and on its surface condition.

The surface condition comprises the properties of the surface border zone. These are chemical, mechanical and geometrical properties, Fig. 1.1.

The chemical and mechanical properties comprise chemical composition, grain, hardness, strength and inhomogeneities. The properties of the surface border zone may be different from those in the core zone.

The geometrical properties are defined as deviations from geometrical ideal elements (features) of the workpiece. Geometrical ideal elements (features) are parts of the entire workpiece surface, which have geometrical, unique and nominal form as e.g. planes, cylinders, spheres, cones and tori. They can also be derived as, for example, axes, section lines, generator lines, lines of highest points and edges, Fig. 1.2. Geometrical deviations are

- size deviations
- form deviations
- orientational deviations
- locational deviations

- waviness
- roughness
- surface discontinuities
- edge deviations

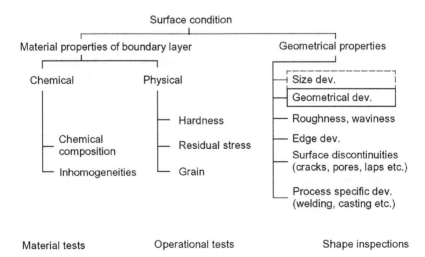

Fig. 1.1 Properties of surfaces and their tests and inspections

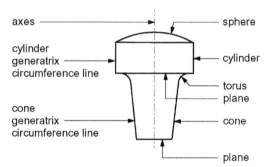

Fig. 1.2 Examples of geometrical elements

Size deviation is the difference between actual size and nominal size. It has to be distinguished between

- deviation of the actual local size from the nominal linear size or from the nominal angular size;
- deviation of the actual substitute size from the nominal substitute size, see 8.

The actual local linear sizes are assessed by two-point-measurements (ISO 8015, ISO 286 and ISO 14 660-2), see Fig. 8.26. The actual local angular sizes are assessed by angular measurements of averaged lines (ISO 8015, ISO 1947), see Fig. 3.45.

The actual local sizes of the same geometrical element of one workpiece are different depending on their location. The actual substitute size is unique and therefore representative for the entire geometrical element of one workpiece.

Size deviations are assessed over the entire geometrical element. They originate mainly by imprecise adjustment of the machine tool and by variations during the manufacturing process, e.g. due to tool wear.

For special functions the new standard ISO 14 405 also presents other definitions of sizes and their drawing modifier symbols.

For example, diameter defined as circumference divided by π with the modifier CC or diameter of the maximum inscribed cylinder with the modifier GX. The two-point size is according to ISO 14 405, the default (without the modifier), and used in this book (in contrast to ASME Y14.5 rule #1).

Form deviation is the deviation of a feature (geometrical element, surface or line) from its nominal form (Fig. 1.3). If not otherwise specified form deviations are assessed over (or along) the entire feature. Form deviations are originated, for example, by the looseness or error in guidances and bearings of the machine tool, deflections of the machine tool or the workpiece, error in the fixture of the workpiece, hardness deflection or wear. The ratio between width and depth of local form deviations is in general more than 1000:1 (VDI/VDE 2601).

Orientational deviation is the deviation of a feature from its nominal form and orientation. The orientation is related to one or more (other) datum feature(s). The orientational deviation includes the form deviation (Fig. 1.3). If not otherwise specified orientational deviations are assessed over the entire feature. Orientational deviations are

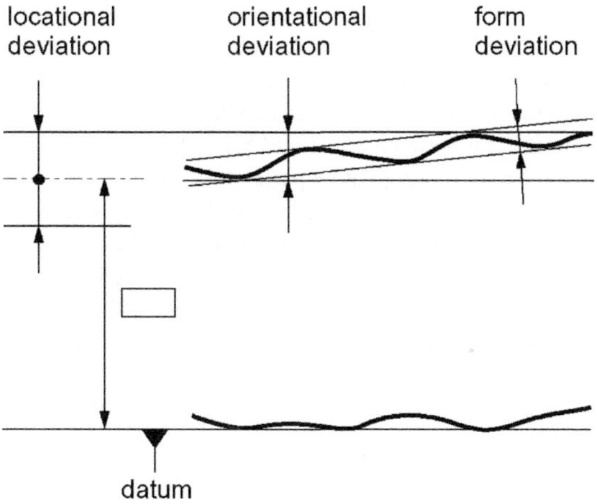

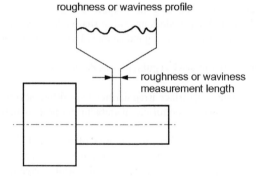

Fig. 1.3 Form deviation, orientational deviation and locational deviation

Fig. 1.4 Assessment of roughness or waviness

originated similarly as form deviations. They are originated also by erroneous fixture of the workpiece after remounting on the machine tool.

Locational deviation is the deviation of a feature (surface, line, point) from its nominal location. The location is related to one or more (other) datum feature(s). The locational deviation includes also the form deviation and the orientational deviation (of the surface, axis, or median face) (Fig. 1.3). If not otherwise specified locational deviations are assessed over the entire feature. Locational deviations are originated similarly as size, form and orientational deviations.

Waviness is mostly more or less periodic irregularities of a workpiece surface with spacings greater than the spacings of its roughness (DIN 4774). The ratio between spacing and depth of the waviness is in general between 1000:1 and 100:1. In general more than one wave can be recognized (VDI/VDE 2601). The waviness is assessed from one or more representative parts of the surface (Fig. 1.4). Waviness is originated by eccentric fixture during the manufacturing process or by form deviations of the cutter or by vibrations of the machine tool and/or the cutting tool and/or the workpiece (DIN 4760). Waviness will not be dealt with in the following.

Roughness is periodic or non-periodic irregularities of a workpiece surface with small spacings inherent of the forming process. The ratio between spacing and depth of the roughness is in general between 150:1 and 5:1 (VDI/VDE 2601). The roughness is assessed from one or more representative parts of the surface (Fig. 1.4).

Roughness is originated by the direct effect of the cutting edges (VDI/VDE 2601), i.e. by imprinting the cutting edges on the surface. Due to the cutting process (tear chip, shear chip) the print will be modified. Other origins are deformations from blasting, gemmation with galvanizing, crystallization, and chemical effects (mordants, corrosion). Roughness ranges down to the crystal structure (DIN 4760). Roughness will not be dealt with in the following.

Surface discontinuity is an isolated imperfection of the surface like a crack, pore or lap. In general it is not taken into account when assessing deviations of size, form, orientation, location, waviness and roughness. The definitions and sizes and permissibilities of surface discontinuities have to be dealt with separately.

At this time there are very few standards on this subject available, e.g. for fasteners ISO 6157 and for hot milled steel products ISO 9443. Surface discontinuities will not be dealt with in the following.

Edge deviations are deviations of the workpiece edge zone from the geometrical ideal shape like burr or abraded edges instead of sharp edges. ISO 13 715 defines tolerances for edges and gives the drawing indications. They are dealt with in clause 22 of this book.

The **classification of surface irregularities** as described above is useful for the following reasons:

(a) The different kinds of surface irregularities have different origins in the manufacturing process. In order to control the manufacturing process these kinds must be assessed separately.

(b) The different kinds of surface irregularities often have different effects on the suitability of the surface for its purpose. For example, on raceways of ball bearings, waviness has a great influence on lifetime and noise while roughness has little influence. In order to specify the permissible function-related deviations, the different kinds must be specified separately.

(c) The depths of the irregularities vary in large ranges, between about $0.1~\mu m$ (and sometimes smaller) with roughness and about $100~\mu m$ (and sometimes more) with form deviations. The ratio between spacing and depth of the irregularities also varies greatly. The smallest ratio between spacing and depth occurs with cracks and is in general smaller than 5:1. Whereas the ratio between spacing and depth with form deviations is in general greater than 1000:1. Because of these wide ranges the requirements for measuring devices and for diagrams are quite different. For the assessment of different kinds of the irregularities (deviations) different kinds of measuring instruments with different magnifications and different profile diagrams with different ratios of horizontal to vertical magnifications are used.

The definitions of the different kinds of irregularities (deviations) are rather uncertain. There are no distinct borderlines. Therefore it was discussed in ISO to define borderlines in terms of defined spacing of irregularities or in terms of defined ratios between spacing and depths of irregularities or in terms of defined ratios between spacing of irregularities and feature lengths. However, it was decided to retain the definitions according to the causes of the irregularities (ISO 4287, ANSI B46.1, BS 1134 and DIN 4760).

There is another distinction between **micro and macro deviations**. Macro deviations are those, that can be assessed with usual measuring devices for the assessment of size, form, orientation and location (e.g. dial indicator). Micro deviations are those that are assessed with roughness or waviness measuring instruments. Macro deviations are assessed over the entire feature length; micro deviations are assessed from a representative part of the surface. Also, there is no distinct borderline because sometimes parts of the waviness will contribute to the result of the measured macro deviations and sometimes parts of the form deviations will contribute to the result of the measured micro deviations (waviness).

Figure 1.5 gives an idea of the combination (superposition) of the kinds of irregularities (deviations) on a surface.

Geometrical deviation Profile diagram	Description Examples of origin
1st order: Form	errors in guidance of machine tool, deflections of machine tool or workpiece, error in fixture of workpiece, warping, wear
2nd order: Waviness	eccentric fixture, form deviation of tool, vibration
3rd order: Roughness 	grooves, form of tool cutting edge, horizontal and vertical feed
4th order: Roughness	cutting process (tear chip, shear chip), deformation from blasting, gemmation with galvanizing
5th order: Roughness not presentable	crystallization process, mordant, corrosion
6th order: Roughness not presentable	crystal structure
Superposition	actual surface

Fig. 1.5 Superposition of surface deviations (DIN 4760)

2

Principles for Tolerancing

It is impossible to manufacture workpieces without deviations from the nominal shape. Workpieces always have deviations of size, form, orientation and location.

When these deviations are too large the usability of the workpiece for its purpose will be impaired. When during manufacturing attempts are made to keep these deviations as small as possible, in order to avoid the impairment of usability, in general the production is too expensive and the product is hard to sell.

In general, competition forces the use of all possibilities for economic production, including possibilities, arising from current developments. Therefore, it is necessary that the drawing tolerances define the workpiece completely, i.e. each property (size, form, orientation and location) must be toleranced. Only then is the manufacturer able to choose the most economic production method, e.g. depending on the number of pieces to be produced and on the production methods available.

Incompletely toleranced drawings result in:

- questions for the production-planning engineer;
- questions for the manufacturing engineer;
- questions for the inspection engineer;
- reworking;
- defects, damages.

Only completely toleranced drawings enable the production of workpieces to be as precise as necessary and as economic as possible. This is necessary for competition.

When all tolerances necessary to define the workpiece completely are indicated individually the drawing becomes overloaded with indications and is hard to read. Therefore general tolerances should be applied.

General tolerances shall be equal to or larger than the customary workshop accuracy. The customary workshop accuracy is equal to those tolerances the workshop does not exceed with normal effort using normal workshop machinery. Larger tolerances bring no gain in manufacturing economy. The normal workshop accuracy depends on the workshop machinery that produces the largest deviations (disregarding exceptions which are to be dealt with in certain cases). The customary workshop accuracy is in general the same within one field of industry. For example, the customary workshop accuracy for material removal in the machine-building industry corresponds to the general tolerances ISO 2768-mH.

When there is no appropriate ISO Standard or National Standard available a company standard should be elaborated.

The general tolerances shall be applied by an indication in or near the title box of the drawing.

Tolerances that must be smaller have to be indicated individually, see 16.

3

Principles for Geometrical Tolerancing

3.1 Symbols

The symbols for the drawing indications of geometrical tolerances according to ISO 1101, ISO 5459, ISO 286 and ISO 10579 are shown in Tables 3.1 and 3.2.

Table 3.2 shows the symbols for the geometrical tolerance characteristics.

With the symbols of the left column of Table 3.2 all kinds of geometrical tolerances can be expressed. The symbols of the right column cover special cases that occur very

Table 3.1 Symbols for geometrical tolerancing (see also 21)

Symbol	Interpretation
3 x // ø 0.02 CZ A	Number (3) of toleranced features when more than 1
	Additional indications (e.g. ø 12 ± 0.1)
	Datum letter
	Symbol for common zone
	Tolerance value
	Symbol for cylindrical or circular tolerance zone
	S ø: spherical tolerance zone
	Symbol for toleranced characteristic
	Additional indications (e.g. MD)
	Arrow line to the toleranced feature
A	Arrow with letter to the toleranced feature
A	Datum triangle and datum box
	Axis or median face as toleranced feature
	Section line, generatrix or surface as toleranced feature or as datum
20	Theoretical exact dimension

Table 3.1 (Cont.)

Symbol	Interpretation

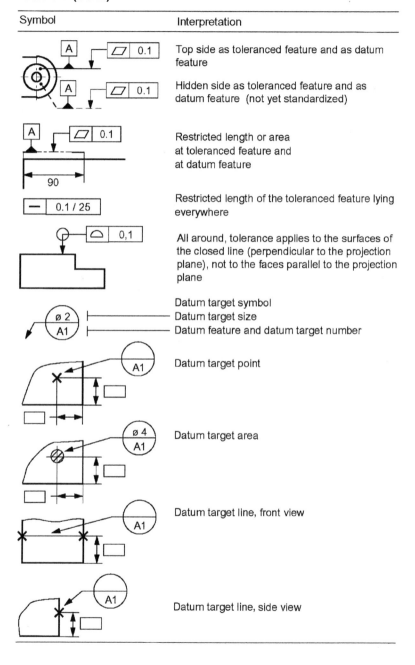

Top side as toleranced feature and as datum feature

Hidden side as toleranced feature and as datum feature (not yet standardized)

Restricted length or area at toleranced feature and at datum feature

Restricted length of the toleranced feature lying everywhere

All around, tolerance applies to the surfaces of the closed line (perpendicular to the projection plane), not to the faces parallel to the projection plane

Datum target symbol
Datum target size
Datum feature and datum target number

Datum target point

Datum target area

Datum target line, front view

Datum target line, side view

Table 3.1 (Cont.)

Symbol	Interpretation
Ⓔ	Envelope requirement
Ⓜ	Maximum material requirement
Ⓛ	Least material requirement
Ⓡ	Reciprocity requirement
Ⓟ	Projected tolerance zone
Ⓕ	Flexible part in free state
Ⓢ	Regardless of feature size (ASME Y14.5)
Ⓣ	Tangent method (ASME Y14.5)
Ⓤ	Unsymmetrical tolerance zone (ASME Y14.8)
MD	Applies to major diameter of threads, splines, gears
PD	Applies to pitch diameter of threads, splines, gears
LD	Applies to (least) minor diameter of threads, splines, gears
LE	Applies to line elements (section lines)
NC	Not convex
ACS	Any cross-section

Further symbols are planned according to ISO 14 405 (in preparation)

frequently. These symbols are standardized in addition, and should be used in order to facilitate readability of drawings.

The proportions of the symbols for the drawing indications are standardized in ISO 7083. However, some CAD systems deviate from this standard. This may be tolerable as long as the drawing indications are unambiguous (as demonstrated in this book).

3.2 Definitions of geometrical tolerances

In order to specify geometrical tolerances, the workpiece is considered to be composed of features (geometrical elements), such as planes, cylinders, cones, spheres, tori, etc. (Fig. 1.2).

Table 3.2 Geometrical tolerances and tolerance symbols

Unrelated geometrical tolerances (form tolerances)				
Profile (form) of lines (line profile)*			⌒	
		Straightness		—
		Roundness		○
Profile (form) of surfaces (surface profile)*			◠	
		Flatness		▱
		Cylindricity		⌀
Related geometrical tolerances				
Orientation	Angularity		∠	
		Parallelism		//
		Perpendicularity		⊥
Location	Position		⊕	
		Coaxiality		◎
		Symmetry		⩵
Run-out	Circular run-out	Circular radial run-out	↗	
		Circular axial run-out		
	Total run-out	Total radial run-out	↗↗	
		Total axial run-out		

* The tolerances of profile of lines and of profile of surfaces may also be related
to datums (Figs 4.2 and 4.4).

Form tolerance is the permitted maximum value of the form deviation (see 18.6 and
18.7.1 to 18.7.7). According to ISO 1101, there are defined form tolerance zones within
which all points of the feature must be contained. Within this zone, the feature may have
any form, if not otherwise specified. The tolerance value defines the width of this zone
(Fig. 1.3).

Form tolerances limit the deviations of a feature from its geometrical ideal line or surface form. Special cases of line forms with special symbols are straightness and roundness (circularity) (Fig. 3.1). Special cases of surface forms with special symbols are
flatness (planarity) and cylindricity (Fig. 3.2).

Orientation tolerance is the permitted maximum value of the orientation deviation
(see 18.6 and 18.7.8). According to ISO 1101, there are defined orientation tolerance
zones within which all points of the feature must be contained. The orientation tolerance
zone is in the geometrical ideal orientation with respect to the datum(s). The tolerance
value defines the width of this zone (Fig. 1.3).

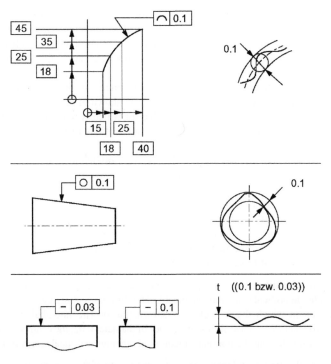

Fig. 3.1 Form tolerances of lines: drawing indications and tolerance zones

Orientation tolerances limit the deviations of a feature from its geometrical ideal orientation with respect to the datum(s). Special cases of orientation with special symbols are parallelism (0°) and perpendicularity (90°) (Fig. 3.3).

The orientation tolerance also limits the form deviation of the toleranced feature (exceptions according to ASME Y14.5, see 21.1.6), but not of the datum feature(s). If necessary, a form tolerance of the datum feature(s) must be specified.

Location tolerance is twice the permitted maximum value of the location deviation (see 18.6 and 18.7.9). According to ISO 1101, there are defined location tolerance zones within which all points of the feature must be contained. The location tolerance zone is in the geometrical ideal orientation and location with respect to the datum(s). The tolerance value defines the width of this zone (Fig. 1.3).

Location tolerances limit the deviations of a feature from its geometrical ideal location (orientation and distance) with respect to the datum(s). Special cases of location with special symbols are coaxiality (when toleranced feature and datum feature are cylindrical) and symmetry (when at least one of the features concerned is prismatic) where the nominal distance between the axis or median plane of the toleranced feature and the axis or median plane of the datum feature is zero (Fig. 3.4).

The location tolerance also limits the orientation deviation and the form deviation of the toleranced feature (plane surface or axis or median face) (exceptions according to ASME Y14.5, see 21.1.5), but not the form deviation of the datum feature(s). If necessary, a form tolerance for the datum feature(s) must be specified.

Run-out tolerances are partly orientation tolerances (axial circular run-out tolerance, axial total run-out tolerance) and partly location tolerances (radial circular run-out tolerance, radial total run-out tolerance). However, according to ISO 1101, they are considered as separate tolerances with separate symbols, because of their special measuring method.

Typical drawing indications and the relevant geometrical tolerance zones of tolerances of form, orientation, location and run-out are shown in Figs 3.1 to 3.11.

Which geometrical tolerances are applicable for which type of feature is shown in Table 3.3. The table also shows the possible combinations of toleranced feature and datum feature.

The possible tolerance zones, their shapes, their orientations and locations, their widths and their lengths are described in 3.3.

The possibilities of specifying datums are described in 3.4.

The definitions of axes and median faces are dealt with in 3.5.

Special rules for screw threads, gears and splines are described in 3.6.

The differences between angularity tolerances according to ISO 1101 and angular dimension tolerances according to ISO 8015 are described in 3.7.

Form tolerances of lines are shown in Fig. 3.1.

Line profile tolerances (Fig. 3.1 top). The nominal (theoretical, geometrical ideal) line is defined by theoretical exact dimensions (TEDs). In each section, parallel to the plane of projection in which the indication is shown, the profile line shall be contained between two equidistant lines enveloping circles of diameter 0.02, the centres of which are located on a line having the nominal (theoretical, geometrical ideal) form (see also 4).

Roundness (circularity) tolerance (Fig. 3.1 centre). In each cross-section of the conical surface the profile (circumference) shall be contained between two coplanar concentric circles with a radial distance of 0.02 (see also Fig. 3.18).

Straightness tolerance (Fig. 3.1 bottom). In each section, parallel to the plane of projection in which the indication is shown, the profile shall be contained between two parallel straight lines 0.03 apart (see also 3.3.1 and Fig. 3.19).

For further examples of form tolerances of lines see 4, 9, 11, 12, 20.1.2, 20.2, 20.3, 20.8 and 20.9.

Form tolerances of surfaces are shown in Fig. 3.2.

Surface profile tolerances (Fig. 3.2 top). The nominal (theoretical, geometrical ideal) surface is defined by theoretical exact dimensions (TEDs). The surface shall be contained between two equidistant surfaces enveloping spheres of diameter 0.03, the centres of which are located on a surface having the nominal (theoretical, geometrical ideal) form (see also 4).

Cylindricity tolerance (Fig. 3.2 centre). The surface shall be contained between two coaxial cylinders with a radial distance 0.05.

Flatness tolerance (Fig. 3.2 bottom). The surface shall be contained between two parallel planes 0.05 apart (see also 20.2).

For tolerancing of cones see 5.

For further examples of form tolerances see 20,1.2, 20.3, 20.8 and 20.9.

Orientation tolerances are shown in Fig. 3.3.

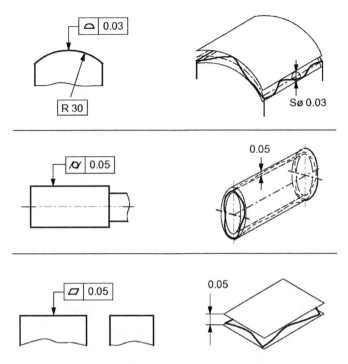

Fig. 3.2 Form tolerances of surfaces: drawing indications and tolerance zones

Angularity tolerance (Fig. 3.3 top). The actual axis shall be contained between two parallel planes 0.1 apart that are inclined at the theoretically exact angle 60° to the datum A.

Perpendicularity tolerance (Fig. 3.3 centre). The surface shall be contained between two parallel planes 0.08 apart that are perpendicular to the datum A.

Parallelism tolerance (Fig. 3.3 bottom). The surface shall be contained between two parallel planes 0.08 apart that are parallel to the datum A.

For further examples of orientation tolerances see 3.3, 20.1.1 and 20.8.

Location tolerances are shown in Fig. 3.4.

Positional tolerances (Fig. 3.4 top). The theoretical exact (nominal) position is defined by the theoretical exact dimensions (TEDs) with respect to the datums A, B and C. The actual axis shall be contained within a cylinder of diameter 0.05, with an axis that coincides with the theoretical exact position.

Coaxiality tolerance (Fig. 3.4 centre). The actual axis shall be contained within a cylinder of diameter 0.03 coaxial with the datum axis A. When the features are practically two dimensional (thin sheet, engraving) the tolerance is also referred to as the concentricity tolerance.

Symmetry tolerance (Fig. 3.4 bottom). The actual median face shall be contained between two parallel planes 0.08 apart that are symmetrically disposed about the datum median plane A.

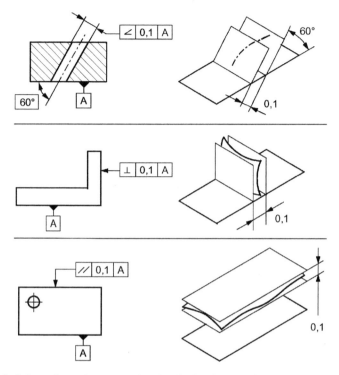

Fig. 3.3 Orientation tolerances: drawing indications and tolerance zones

For further examples of location tolerances see 3.3, 6, 7, 9, 11, 19 and 20.

Radial run-out tolerances are shown in Fig. 3.5.

Circular radial run-out (Fig. 3.5 top). In each plane perpendicular to the common datum axis A–B the profile (circumference) shall be contained between two circles concentric with the datum axis A–B and with a radial distance of 0.1.

Total radial run-out tolerance (Fig. 3.5 bottom). The surface shall be contained between two cylinders coaxial with the datum axis A–B and with a radial distance of 0.1.

During checking of the circular radial run-out deviation, the positions of the dial indicator are independent of each other. However, during checking of the total radial run-out deviation, the positions of the dial indicator are along a guiding (straight) line parallel to the datum axis A–B (see 18.7.8.3).

Therefore the straightness deviations and the parallelism deviations of the generator lines of the toleranced cylindrical surface are limited by the total radial run-out tolerance, but not by the circular radial run-out tolerance (Fig. 3.6).

See also Fig. 20.59.

Axial run-out tolerances are shown in Fig. 3.7.

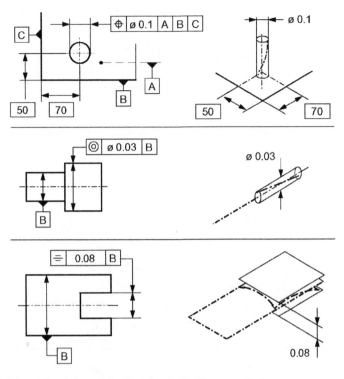

Fig. 3.4 Location tolerances: drawing indications and tolerance zones

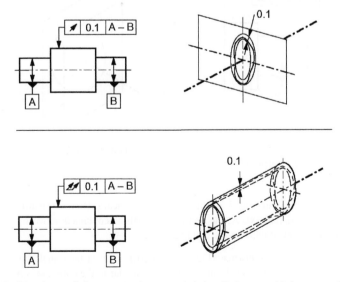

Fig. 3.5 Circular radial run-out tolerance, total radial run-out tolerance: drawing indications and tolerance zones

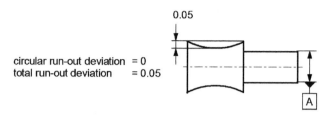

circular run-out deviation = 0
total run-out deviation = 0.05

Fig. 3.6 Workpiece with total radial run-out deviation, but without circular radial run-out deviation

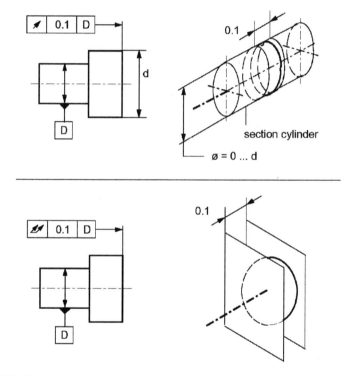

Fig. 3.7 Circular axial run-out tolerance, total axial run-out tolerance: drawing indications and tolerance zones

Circular axial run-out (Fig. 3.7 top). In each cylindrical section (measuring cylinder) coaxial with the datum axis A, the section line shall be contained between two circles 0.1 apart and perpendicular to the datum axis A.

Total axial run-out tolerance (Fig. 3.7 bottom). The surface shall be contained between two parallel planes 0.1 apart and perpendicular to the datum axis A.

During checking of the circular axial run-out deviation, the positions of the dial indicator are independent of each other. However, during checking of the total axial run-out

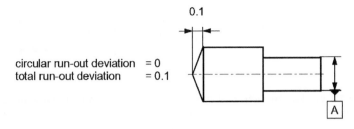

Fig. 3.8 Workpiece with total axial run-out deviation, but without circular axial run-out deviation

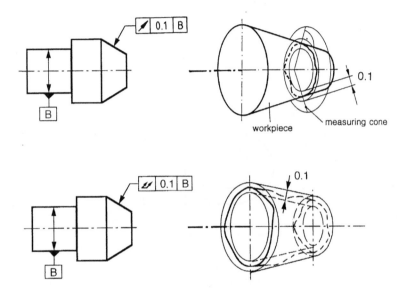

Fig. 3.9 Circular run-out tolerance in any direction, total run-out tolerance in any direction: drawing indications and tolerance zones

deviation, the positions of the dial indicator are along a guiding (straight) line perpendicular to the datum axis A (see 18.7.8.3).

Therefore the flatness deviations of the toleranced surface are limited by the total axial run-out tolerance, but not by the circular axial run-out tolerance (Fig. 3.8).

Run-out tolerances in any direction are shown in Fig. 3.9.

Circular run-out tolerance in any direction (Fig. 3.9 top). In each conical section (measuring cone) coaxial with the datum axis B and perpendicular to the nominal toleranced surface (defining the measuring cone angle) the section line shall be contained between two circles 0.1 apart and perpendicular to the datum axis B.

Total run-out tolerance in any direction (Fig. 3.9 bottom). The surface shall be contained between two cones coaxial with the datum axis B and with a radial distance of 0.1 (measured perpendicular to the nominal cone surfaces) (see also Fig. 5.4).

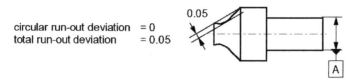

circular run-out deviation = 0
total run-out deviation   = 0.05

Fig. 3.10 Workpiece with total run-out deviation, but without circular run-out deviation

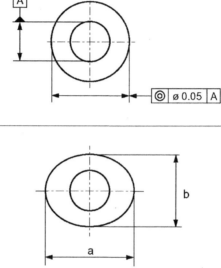

coaxiality deviation ≈ 0
run-out deviation   = (a − b)/2

Fig. 3.11 Workpiece with radial run-out deviation, but practically without coaxiality deviation

During checking of the circular run-out deviation in any direction, the positions of the dial indicator are independent of each other. However, during checking of the total run-out deviation in any direction, the positions of the dial indicator are along a guiding line (theoretical exact generator line of the toleranced future) parallel to its theoretical exact position with respect to the datum axis B.

Therefore the deviations of the generator line of the toleranced feature are limited by the total run-out tolerance in any direction, but not by the circular run-out tolerance in any direction (3.10).

Coaxiality tolerance and **radial run-out tolerance** are different. The coaxiality tolerance assesses the deviation of the axis from the datum axis, while the radial run-out tolerance assesses the deviation of the circumference line from a coaxial circle. The radial run-out deviation is composed of the coaxiality deviation and parts of the roundness deviation. For limiting unbalance, for example, the coaxiality tolerance is appropriate to

Table 3.3 Possible tolerancing of features (form elements)

Tolerance		Symbol	Toleranced feature					Datum feature				
			Section line	Edge	Axis	Median face	Surface	Section line	Edge	Axis	Median face	Surface
Unrelated tolerance	Line profile	⌒	x	x	x							
	Straightness	–	x	x	x							
	Roundness	O	x	x								
	Surface profile	⌒				x	x					
	Flatness	▱				x	x					
	Cylindricity	⌀⁄					x					
Related tolerance	Line profile with datum	⌒	x	x				x*	x	x	x	x
	Surface profile with datum	⌒					x		x	x	x	x
	Inclination	∠	x*	x	x		x	x*	x	x	x	x
	Parallelism	//	x*	x	x	x	x	x*	x	x	x	x
	Perpendicularity	⊥	x*	x	x	x	x	x*	x	x	x	x
	Position	⊕		x	x	x	x		x	x	x	x
	Coaxiality	◎			x					x		
	Symmetry	⚌			x	x				x	x	
	Circular run-out	↗	x	x						x		
	Total run-out	↗↗		x		x				x		
* Indication necessary: "lines" or "cross-section lines only"												

the function rather than the radial run-out tolerance. Figure 3.11 shows an example where there is a radial run-out deviation, but practically no coaxiality deviation.

Possibilities of geometrical tolerancing of features are listed in Table 3.3, which indicates the possible types of tolerancing depending on the type of feature.

The drawing indication "lines" may be omitted when the drawing already shows that only section lines can be applied (Figs 20.7 and 20.9). See also 20.2.

Further possibilities are positional tolerancing of the centre point of a sphere or positional tolerancing or coaxiality (concentricity, same symbol as coaxiality) tolerancing of the centre point within a cross-section. These points may also be used as a datum.

Limitation of deviations by geometrical tolerances of higher order is shown in Fig. 1.3.

Related geometrical tolerances (tolerances of orientation and location according to ISO 1101 and ASME Y14.5M) define zones within which all points of the toleranced feature have to be contained. Therefore related geometrical tolerances also limit certain form deviations. Locational tolerances also limit the orientational deviations and the form deviations of the toleranced feature (surface or line). (Exceptions according to ASME Y14.5: orientation tolerances for axes and median planes and positional tolerances are applied to the (straight) axes or (plane) median planes of mating features and do not limit the form deviations of the actual feature axes or median surfaces, see 21.1.3 and 21.1.6.)

For example, related geometrical tolerances of surfaces also limit the form deviations of these surfaces (Fig. 3.3 centre and bottom, Fig. 3.5 bottom and Fig. 3.7 bottom).

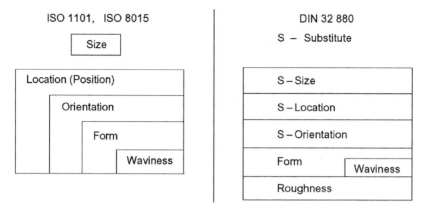

Fig. 3.12 Assessment of types of geometrical deviations by geometrical tolerances

Related geometrical tolerances of section lines limit the form deviations of the section lines but not the form deviations of the surface (in the other direction) (Fig. 3.5 top and Fig. 3.7 top).

Related geometrical tolerances of axes or median faces limit the form deviations of the axes or median faces, but not of the pertinent surfaces (Fig. 3.3 top and Fig. 3.4). (According to ISO 1101 for all geometrical tolerances and according to ASME Y14.5M for all tolerances with the exception of positional tolerances and orientation tolerances, it is assumed that actual axes and actual median faces of workpieces are subject to form deviations (from the geometrical ideal shape) (see also 3.5).)

Locational tolerances (e.g. positional tolerances) limit the location and also the orientation and form of the toleranced feature (e.g. axis, planar surface) (Fig. 3.12, see also 8 and 18.10).

The indication of form tolerances is not necessary when the related geometrical tolerance already limits the form deviations to a sufficient extent.

According to the former East European Standards (former Comecon Standards), form deviations are not limited by related geometrical tolerances (see 21.4).

The draft standard DIN 32 880 will be substituted by a future ISO Standard on VD&T.

3.3 Tolerance zone

3.3.1 Form of the tolerance zone

Depending on the toleranced characteristic and depending on the drawing indication the tolerance zone is one of the following:

- area within a circle (Fig. 3.13);
- area between two concentric circles (Fig. 3.14);
- area between two equidistant lines or between two parallel straight lines (Fig. 3.15);
- space within a sphere (Fig. 3.16);
- space within a cylinder (Figs 3.17 and 3.20);

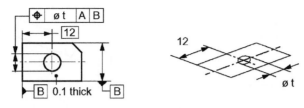

Fig. 3.13 Circular tolerance zone

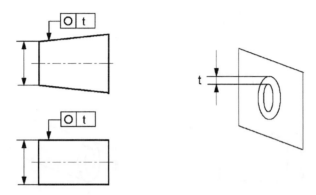

Fig. 3.14 Tolerance zone between two concentric circles

Fig. 3.15 Tolerance zone between two parallel straight lines in a plane

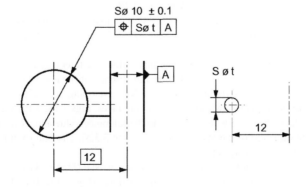

Fig. 3.16 Spherical tolerance zone

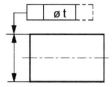
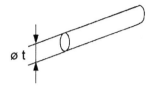

Fig. 3.17 Cylindrical tolerance zone

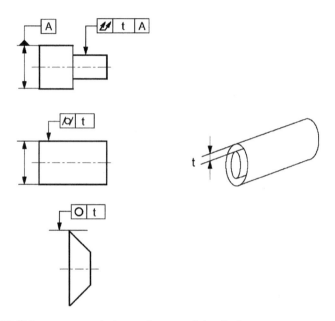

Fig. 3.18 Tolerance zone between two coaxial cylinders

- space between two coaxial cylinders (Fig. 3.18);
- space between two equidistant faces or between two parallel planes (Fig. 3.19);
- space within a parallelepiped (Figs 3.20 and 3.21; see also text to Fig. 6.2).

When the ø symbol precedes the tolerance value, the tolerance zone is cylindrical (Figs 3.17, 3.20) or circular (Fig. 3.13). In the absence of the ø symbol, the tolerance zone is limited by two lines (straight lines, circles, enveloping lines; Figs 3.14 and 3.15) or by two surfaces (planes, cylinders, enveloping faces; Figs 3.18, 3.19, 3.20, 3.21 and 3.2), the distance between which is equal to the tolerance value and, according to ISO 1101, in the direction of the leader line arrow (see 3.3.3). The unit (mm or in) of the dimensions in the drawing applies.

For section lines the tolerance zones are two dimensional (Figs 3.14 and 3.15). For thin parts the tolerance zone is practically two dimensional (Fig. 3.13). For edges, axes and median faces and surfaces the tolerance zone is three dimensional (Figs 3.17 to 3.21).

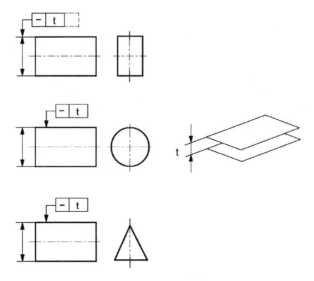

Fig. 3.19 Tolerance zone between two parallel planes

For the straightness tolerance of generator lines of cylindrical or conical features the standards do not specify whether section lines in planes containing the axis* or lines of the highest points are applied. Often the function refers to the lines of the highest points. Therefore they are applied here (Fig. 3.19). In case of doubt the indication "high point line" or "section line" is recommended.

For axes the tolerance zone is three dimensional. When the tolerance is indicated in only one direction, the deviation is limited in only this direction (Fig. 3.20 top)**. When the tolerance value is preceded by the ø symbol the tolerance zone is cylindrical (Fig. 3.20 bottom). When the tolerances are indicated in two directions perpendicular to each other for axes, the tolerance zone is a parallelepiped (Figs 3.20 centre, 3.21 and 3.24). When the tolerance for an axis is indicated in two directions defined by two datums independent of each other (Fig. 3.23), the tolerance zone does not necessarily have right angles.

3.3.2 Location and orientation of the tolerance zone

With **locational tolerances** and with **run-out tolerances**, the tolerance zone is located centrally or coaxially or symmetrically with respect to the theoretical exact (nominal) location (Figs 3.4 to 3.9).

With **orientation tolerances**, the tolerance zone is orientated with respect to the datum. The distance from the datum is not limited.

---

* Axis of the least squares cylinder (default according to ISO 14 660-2) or cone or axis of the contacting cylinder or cone directed according to the minimum rock requirement or axis of the pair of coaxial cylinders or cones with least distance (Chebyshev) (see 3.5).

** In ISO 1101:1983 the tolerance zone was sometimes shown as a zone projected on- to a plane. According to Clause 13.3 of the standard, there is no difference in the meaning of tolerance zones shown projected onto a plane or shown three dimensionally (Fig. 3.22). In the new edition of ISO 1101:2004 the spatial (three-dimensional) tolerance zones are always shown.

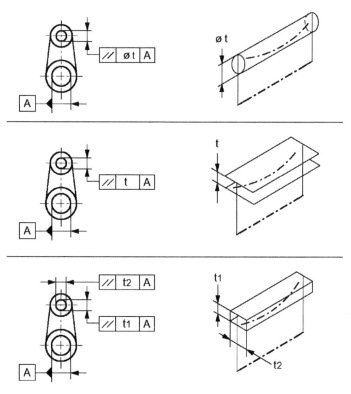

Fig. 3.20 Spatial tolerance zone of a line

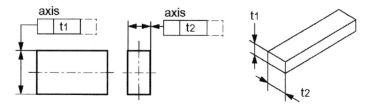

Fig. 3.21 Parallelepipedal tolerance zone

With **form tolerances**, the distance and the orientation of the tolerance zone with respect to other features (form elements) is not limited.

Exceptions are tolerance zones of profile tolerances where the nominal (geometrical ideal) form is defined by theoretical exact dimensions related to datums (Fig. 4.2). In these cases the tolerance zone is symmetrically placed (equidistant) with respect to the nominal form (line or face) in the nominal (theoretical exact) location and orientation with respect to the datum(s). The ways of defining the tolerance zone are imaginary circles or spheres. Their centres are located on the nominal form. The envelopes (tangential) on the circles or spheres form the tolerance zone (Figs 3.1 top, 3.2 top and 2.10).

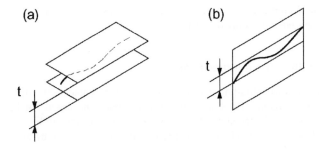

Fig. 3.22 Tolerance zone of axes: (a) spatial representation, (b) projection of the spatial zone onto a plane (without any change of meaning)

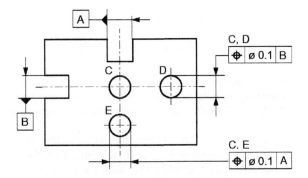

Fig. 3.23 Tolerance zone not necessarily right angular located; tolerance zone of C not necessarily of right angles in the section

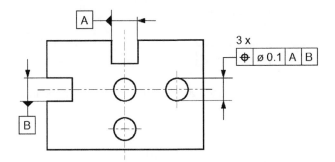

Fig. 3.24 Tolerance zones right angular located; tolerance zone of the centre hole of right angles in the section

Unilaterally disposed tolerance zones (on one side of the nominal form only) or not equally disposed tolerance zones are not provided with ISO 1101. In all cases the theoretical exact dimensions shall be chosen in such a way that the tolerance zone limits are equidistant to the nominal form. This avoids the sometimes difficult estimation (in cases of curved forms) of the geometrical ideal form, following the tolerance zone centres during manufacturing and inspection. According to ASME Y14.5M the tolerance zone may be disposed unilaterally or unequally (see Table 21.2).

3.3.3 Width of the tolerance zone

Tolerance zones have limiting lines or surfaces that are equidistant from the nominal (geometrical ideal) form. The shape of the tolerance zone is independent of the size (diameter or distance) of the feature(s). Exceptions are tolerance zones of line profile or surface profile where the nominal (geometrical ideal) form is defined by theoretical exact dimensions (TEDs).

The width of the tolerance zone (in measuring direction) is equal to the tolerance value, and is directed according to ISO 1101 in the direction of the leader line arrow:

- with tolerances of axes and median faces perpendicular to the toleranced axis or median face;
- with tolerances of surfaces or lines perpendicular to the surface (Fig. 3.25).

An exception is the tolerance zone of roundness, where, if not otherwise specified, the width is defined perpendicular to the axis. (However, with run-out tolerances, if not otherwise specified, the width of the tolerance zone is defined perpendicular to the surface; Fig. 3.25).

When, in exceptional cases (e.g. with cones or curves), the width of the tolerance zone is not perpendicular to the surface or the axis or the median face, the direction of the width of the tolerance zone shall be indicated as shown in Fig. 3.26.

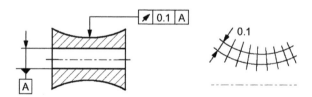

Fig. 3.25 Width of the tolerance zone in the direction of the leader line arrow perpendicular to the surface

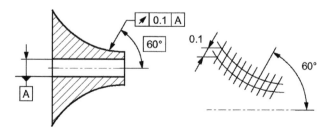

Fig. 3.26 Indicated direction of the width of the tolerance zone

3.3.4 Length of the tolerance zone

If not otherwise indicated, the geometrical tolerance (tolerance zone) applies to the entire length or surface of the feature.

When the leader line arrow points to a thick chain line or area within a thick chain line (Fig. 3.27), the tolerance applies to this region. When the line designating the area within a thick chain line is dashed, the indication applies to the hidden surface (Fig. 3.27c, not yet standardized in ISO 1101).

When the tolerance value is followed by an oblique stroke and another value (e.g. 0.05/100), the second value (100 here) indicates the length within which the tolerance (0.05) applies. The tolerance zone (of 0.05 width) of the specified length (100) applies to all possible locations on the feature. Therefore these tolerances (0.05) can accumulate along the feature (Fig. 3.28).

3.3.5 Common tolerance zone

Isolated (separate) tolerance zones having the same value applied to different features may be indicated as shown in Fig. 3.29.

When the features are to be contained in a common tolerance zone, this is to be indicated by "CZ" after the tolerance value in the tolerance frame (Figs 3.30 and 20.41).*

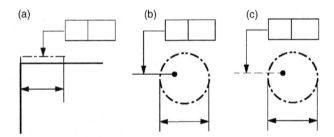

Fig. 3.27 Tolerance zone applied to a restricted region: (a), (b) on the surface shown, (c) on the hidden surface

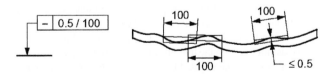

Fig. 3.28 Tolerance zones applied to restricted lengths in any location

* The former practice was, according to ISO 1101:1983, to indicate "common zone" above the tolerance frame.

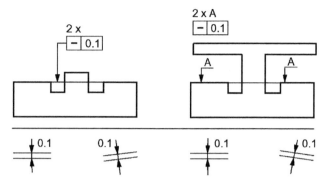

Fig. 3.29 Simplified indication of tolerances that are similar but independent of each other

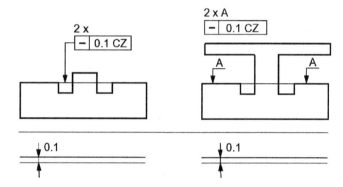

Fig. 3.30 Indication of a common tolerance zone

3.4 Datums

Datums define the orientation and/or location of the tolerance zone. For this purpose, the form deviations of the datum feature(s) are eliminated by the minimum rock requirement (see 18.3.2) or by the indication of datum targets (Fig. 3.36).

The datum may be established by

- one single datum feature;
- two or more datum features of the same priority as a common datum (e.g. a common axis, Fig. 3.31), indicated by a hyphen between the datum letters in the tolerance frame;
- two or more datum features with different but not specified priorities (indicated by a sequence of datum letters in the tolerance frame without separation) (Fig. 3.32); the application of this should be avoided because this indication is not unequivocal and may lead to different measuring results for the same workpiece (Fig. 3.35); this indication has therefore been eliminated from ISO 1101 (now labelled as former practice);
- datum features of different priority (e.g. three-plane datum system according to ISO 5459) (Figs 3.33, 3.34 and 3.35);
- datum targets of different priority on different features (Fig. 3.36).

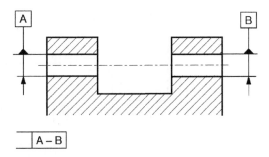

Fig. 3.31 Datums of same priority as common datums

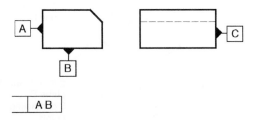

Fig. 3.32 Datums with priorities not specified

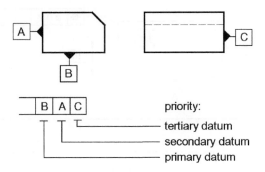

Fig. 3.33 Datums with different priorities

Datums can have different priorities (order of precedence) as follows (Figs 3.34 and 3.35):

- **primary datum**, datum feature orientated according to the minimum rock requirement (see 18.3.2) relative to the simulated datum feature (auxiliary datum element) (see 18.2);
- **secondary datum**, datum feature orientated without tilting relative to the primary simulated datum feature (only by translation and rotation) according to the minimum rock requirement relative to the secondary simulated datum feature.

The secondary simulated datum feature is perpendicular to the primary simulated datum feature.

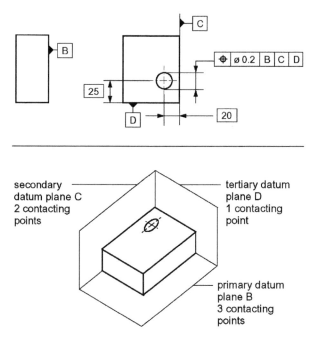

Fig. 3.34 Three-plane datum system according to ISO 5459

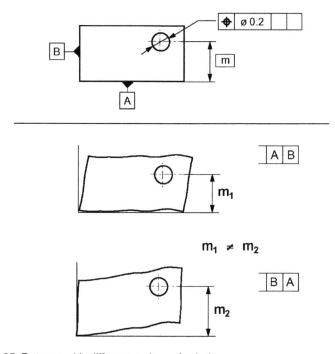

Fig. 3.35 Datums with different orders of priority

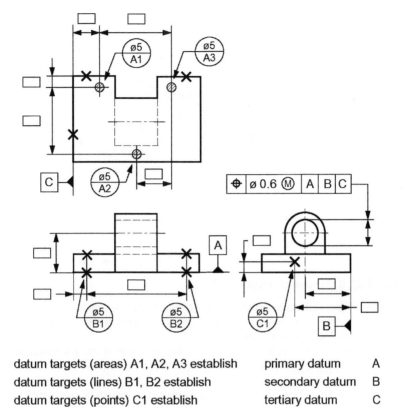

datum targets (areas) A1, A2, A3 establish	primary datum	A
datum targets (lines) B1, B2 establish	secondary datum	B
datum targets (points) C1 establish	tertiary datum	C

Fig. 3.36 Datum targets: planar surfaces

- **tertiary datum**, datum feature positioned without tilting and rotation relative to the primary and to the secondary simulated datum feature only by translation until contacting with the tertiary simulated datum feature.

  The tertiary simulated datum feature is perpendicular to the primary and to the secondary simulated datum feature.

The effect of different orders of priority is illustrated in Fig. 3.35.

In all cases only the outer extremities of the datum features contact the simulated datum features.

Form deviations of the datum features are not limited by the indication as a datum. Therefore, if the general form tolerance (see 16) is not sufficient,* individual form tolerances must be indicated on the datum feature or datum targets according to ISO 5459 must be applied (Fig. 3.36).

---

* A widely used practice is to specify a form tolerance to the datum feature that is not larger than 20% of the geometrical tolerance related to this datum. Then in most cases the workpiece does not rock significantly and the alignment of the workpiece according to the minimum rock requirement is not necessary. However, in extreme cases depending on the shape of the workpiece and on the form deviation, the workpiece still rocks causing measurement errors up to ±50% and more.

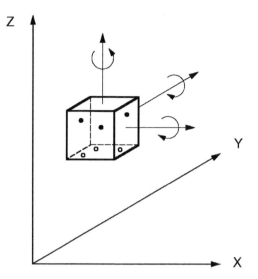

Fig. 3.37 6 degrees of freedom

When datum features have relatively large form deviations compared with the tolerance of orientation or location (e.g. with castings), **datum targets** according to ISO 5459 may be indicated (Fig. 3.36).

A datum target may be a restricted area, a line or a point of specified location (see Table 3.1). They represent the supports of the workpiece during inspection (and manufacturing).

For planar datum features the following datum targets have to be specified:

- at the primary datum   3 targets;
- at the secondary datum   2 targets;
- at the tertiary datum   1 target.

The total of datum targets is 6 according to the total of degrees of freedom (translation along the x, y and z axis, rotation around the x, y and z axis, Fig. 3.37) (each datum target removes one degree of freedom). Table 3.4 shows the necessary number of datum targets for the primary datum depending on its shape.

According to ASME Y14.5M instead of 2 or 4 datum targets on a cylindrical datum feature there may be specified one or two sets of 3 datum targets spaced at 120°. This is for centring devices (three-jaw chuck) used to establish one or two points of the datum axis, the device capable of moving radially at an equal rate from a common axis (see Fig. 3.38; exception of the rule of 6 given above).

The datum targets define a coordinate system; therefore the location of the targets in respect to each other must be dimensioned by theoretical exact dimensions (in order to establish a unique workpiece coordinate system). In contrast to ASME Y14.5M ISO 5459 – 1981 does not yet give this information but will probably be changed in the same way (see Fig. 3.36).

Exceptions of the rules given above are datum features where the datum letter in the tolerance frame is followed by the symbols Ⓜ or Ⓛ (see 9.2 and 9.3.3; 11.2 and 11.3.3).

Table 3.4 Number of datum targets of primary datums

Plane	3
Sphere	3
Cylinder	4
Cone	5
Torus	5

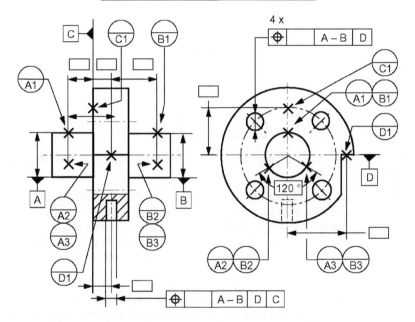

Fig. 3.38 Datum targets: cylindrical surface and three-jaw chuck

Datum features of too short length create great measuring uncertainty. Therefore these short features should not serve as datums. If necessary, the drawing must be altered (Fig. 19.13).

3.5 Axes and median faces

When the leader line arrow is indicated in line with the dimension line, the geometrical tolerance applies to the axis or median face (Fig. 3.39). The form deviations of the generator lines or surfaces are not limited; they may have larger straightness or flatness deviations. When the leader line arrow is indicated on the centre line in the drawing, the geometrical tolerance applies to all axes or median faces that the centre line represents. (The latter indication should be avoided because ISO 1101 does not specify whether a common tolerance zone is required or not.)

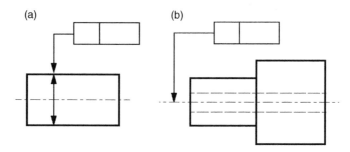

Fig. 3.39 Tolerance indication for axes and median faces: (a) axis or median face; (b) all axes or median faces

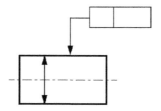

Fig. 3.40 Tolerance indication for surfaces, generator lines and circumference lines

When the leader line arrow is indicated apart from the dimension line (the separation should be at least 4 mm), the geometrical tolerance applies directly to the surface, generator line or circumference line (Fig. 3.40). This indication limits also the straightness deviation of the axis, since the axis cannot deviate more from straightness than the generator lines (Fig. 3.41). A similar consideration applies to the flatness deviation of the median face.

A similar symbolization applies to datums. When the datum triangle and the dimension line are aligned, the axis or median face is the datum. When the datum triangle is indicated apart from the dimension line, the minimum rock requirement applies to the surface or to the generator lines.

When the datum triangle is indicated on the centre line in the drawing, all axes or median faces that the centre line represents apply as a common datum. This is the axis of the smallest cylinder containing all axes or the median plane of two parallel planes of least distance containing all median faces. According to ISO 1101 this indication should be avoided because of the difficulties that arise with inspection.

With axes it is necessary to distinguish between nominal axes, actual axes and datum axes.

Nominal axis is the geometrical ideal (straight) axis of the geometrical ideal feature of nominal size defined in the drawing.

Actual axis is the irregular axis that follows the form deviations of the feature. The following definitions are possible. The axis is composed of the centres of

(a) least squares circles (Gauss),
(b) minimum zone circles (Chebyshev),
(c) contacting circles,

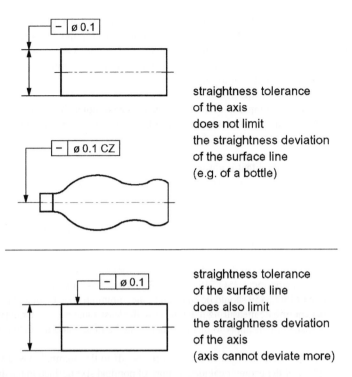

straightness tolerance
of the axis
does not limit
the straightness deviation
of the surface line
(e.g. of a bottle)

straightness tolerance
of the surface line
does also limit
the straightness deviation
of the axis
(axis cannot deviate more)

Fig. 3.41 Straightness tolerance of axes and generator lines

in cross-sections perpendicular to the axis of the

(a) least squares cylinder or cone (Gauss),
(b) minimum zone cylinders or cone (Chebyshev),
(c) contacting cylinder or cone.

The definitions of these circles and cylinders are given in 18.7.2.1 and 18.7.5.1. In the case of a cone the least squares (substitute) cone is the best-fit cone, i.e. the (substitute) cone angle deviates from the nominal cone angle (see 8.3.1, Fig. 8.22), if the cone angle is not indicated as a TED.

The least squares circles and cylinders or cones have advantages in measurement (least number of necessary measured points and least measuring time). The least squares circles are always unique (for all types of roundness deviations). This does not always apply to the minimum zone circles and to the contacting circles.

Minimum zone circles and cylinders and cones have the advantage of being in harmony with long existing specifications in ISO 1101.

Contacting circles and cylinders and cones have the advantage of relating to those parts of the workpiece surface that are in contact with the counterpart in cases of fits. But they have the disadvantage of relating to just a few points of the surface, and therefore are rather unsuitable for assessing the dynamic behaviour (e.g. imbalance) of the workpiece.

In the past it was thought to be unnecessary to define the actual axis. But recent investigations reveal that considerable differences (up to 100%) in the measurement result can arise owing to these differences in the definitions. Therefore it is necessary, especially for the programming of coordinate measuring machines and of form measuring instruments, to standardize definitions unequivocally.

ISO 14 660-2 defines the default as follows.

The actual axis is composed of the least squares centre points in section planes perpendicular to the least squares substitute cylinder or cone (see 8.3.1 and Fig. 8.26).

This definition has been chosen because

- it is the only method that is always unique;
- all assessed points of the workpiece contribute to the determination of the axis (in cases of fits and of material thickness the boundary Ⓔ, Ⓜ, Ⓛ is of interest rather than the axis);
- it needs the least number of points to become sufficiently stable (to meet the required measurement uncertainty).

Datum axis is the axis of the contacting cylinder or cone orientated according to the minimum rock requirement, ISO 5459 (see 18.3.4). Here it is under discussion whether the definition of ISO 5459 should be changed to the minimum zone cylinder or cone (at present most coordinate measuring machines use the least squares cylinder definition).

Similar considerations apply to median planes, median faces and the datum median plane.

Median plane (nominal median plane) corresponds to the nominal axis, and is the symmetry plane of the geometrical ideal feature of nominal size defined in the drawing.

Median face (actual median face) corresponds to the actual axis; e.g. in the case of the least squares method the median face is composed of the centre points between opposite points on opposite surfaces, while the connection of each pair of points is perpendicular to the least squares substitute median plane (median plane between the substitute planes of the two opposite surfaces) (see 8.3.1 and Fig. 8.25).

Datum median plane corresponds to the datum axis, and is the median (symmetry) plane of the contacting pair of parallel planes orientated according to the minimum rock requirement, ISO 5459. Again here it is under discussion whether the definition of ISO 5459 should be changed to the symmetry plane of the pair of minimum zone planes.

Candidate datum method is a substitute for the minimum rock requirement, according to ASME Y14.5 (see 18.3.4).

3.6 Screw threads, gears and splines

Tolerances of orientation or location and datum references specified (indicated) for screw threads apply to the axis of the thread, derived from the pitch cylinder, if not otherwise specified (Figs 3.42 and 3.43).

Tolerances of orientation or location and datum references specified (indicated) for gears and splines must designate the specific feature of the gear or spline to which they apply (e.g. pitch diameter (PD), or major diameter (MD) or minor (least) diameter (LD)).

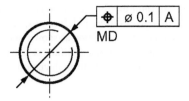

Fig. 3.42 Positional tolerance specified for the major diameter cylinder axis of a thread

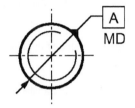

Fig. 3.43 Datum specified as the major diameter cylinder axis of a thread

3.7 Angularity tolerances and angular dimension tolerances

Angular dimension tolerances are usually specified in angular grades or minutes following the nominal angular size (e.g. 45° ± 30′). Angular dimension tolerances define ranges of angles (e.g. 44° 30′ to 45° 30′) for the general direction of actual lines or line elements of surfaces (contacting lines in section planes).

The general direction of the contacting line derived from the actual surface is the direction of the contacting line of geometric ideal form (straight line). The maximum distance of the actual line from the contacting line is the least possible, i.e. the contacting line (e.g. leg of the angle measuring instrument, protractor) contacts the actual surface on the highest point(s) in an average direction (Fig. 3.44). Therefore the angular

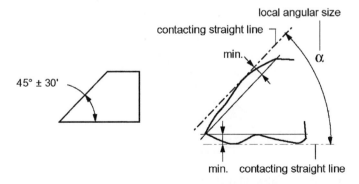

Fig. 3.44 Angular dimension tolerance and actual angular size

dimension tolerance does not limit the form deviations of the lines or surfaces consti-
tuting the angle. ASME Y14.5 gives different definitions, see 21.1.2.

Angularity (inclination) tolerances, specified as shown in Fig. 3.45, define tolerance
zones within which all points of the toleranced surface or line must be contained. Therefore the inclination tolerance limits also the form deviations of the toleranced feature.

3.8 Twist tolerance

Some national standards define a twist tolerance instead of the flatness tolerance (e.g.
the German Standards on aluminium sections DIN 1748 and DIN 176 215). The definition is as follows. The section is laid on a flat ground. One end shall contact the
ground as perfectly as possible (minimum rock requirement). At the other end the twist
deviation is measured as the maximum distance from the ground (Fig. 3.46). The twist
tolerance is the maximum permissible value of the twist deviation.

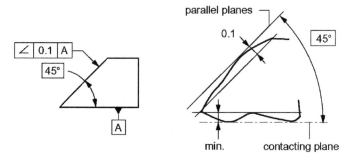

Fig. 3.45 Angularity (inclination) tolerance and tolerance zone

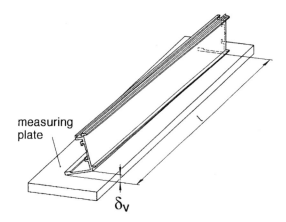

Fig. 3.46 Twist deviation of a section

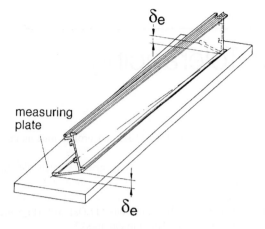

Fig. 3.47 Flatness deviation of a section

At the same workpiece the twist deviation δ_v is larger than the flatness deviation δ_e because for the assessment of the flatness deviation the entire surface of the section (not only one end) must follow the minimum requirement (relative to the flat ground) (Fig. 3.47).

4

Profile Tolerancing

Profile tolerances are defined in ISO 1101, and profile tolerancing is described in ISO 1660.

With profile tolerancing a distinction must be made between tolerancing of the form of lines (symbol ⌒) and of the form of surfaces (symbol ⌓).

The nominal (theoretical exact, geometrical ideal, true) form is to be defined by theoretical exact (rectangular framed) dimensions (TEDs) with (Figs 4.2 and 20.12) or without (Figs 4.1, 4.3 and 20.10) relation to datum(s).*

The tolerance zone of a profile tolerance is defined by (tangential) envelopes on circles (profile tolerance of a line) or on spheres (profile tolerance of a surface) whose diameters are equal to the tolerance value and centred on the nominal form (Figs 4.1, 4.2 and 3.2). Therefore the zone is equally disposed on either side of the nominal profile.

In order to change the location of the tolerance zone the nominal dimensions can be recalculated or the tolerance zone can be unequally disposed relative to the nominal profile as shown in Fig. 4.3.

When the width of the tolerance zone varies within the profile see Fig. 20.10.

The form of the envelope of the circles or spheres (tolerance zone) between the specified points is not standardized. Clearly, the envelope shall alter its form in a smooth manner. Programs for CAD (Computer Aided Design) and CMM (Coordinate Measuring Machines) usually use splines.

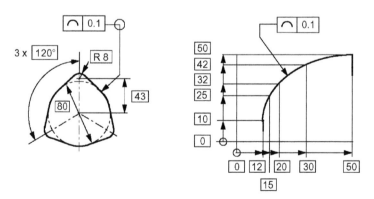

Fig. 4.1 Profile tolerance of lines; the nominal form is defined by consecutive straight lines and circles defined by theoretical exact dimensions without reference to datums

---

* With plain geometrical forms (straightness, roundness, flatness and cylindricity), the drawing defines the nominal form without theoretical exact dimensions. Then, however, the use of the special symbols ⁻, ○, ⟋, ⟋ is recommended.

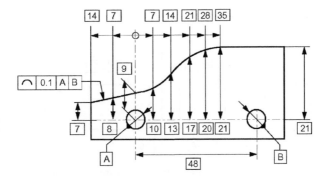

Fig. 4.2 Profile tolerance of lines; the nominal form is defined by theoretical exact dimensions with reference to datums (A–B)

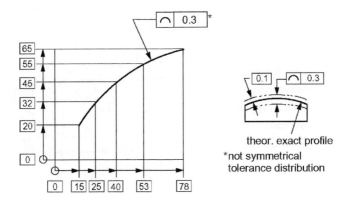

theor. exact profile
*not symmetrical
tolerance distribution

Fig. 4.3 Unequally disposed profile tolerance zone

Profile tolerancing can be used for tolerancing of form, orientation, location and form with size, depending on the indication of datums and theoretical exact dimensions (Fig. 4.4).

Figure 4.5 shows a drawing with profile tolerancing. The holes and the radius are in functional relation with the datum surfaces A, B and C. The drawing shows directly the functional requirements.

Figure 4.6 shows the drawing for the same purpose with dimensional coordinate tolerancing (± tolerancing). Since there is tolerance accumulation and since there is no datum system indicated, it is difficult to assess what are the extreme permissible geometrical deviations of the workpiece contour. Because of the tolerance accumulation and because it is impossible to specify cylindrical tolerance zones with ± tolerances, the tolerances are smaller than with profile tolerancing (and positional tolerancing) and production will become more expensive. Even then it is mostly not possible to indicate the functional requirements in a proper way, i.e. the drawing is wrong.

Profile tolerancing (together with positional tolerancing, maximum material requirement, least material requirement and projected tolerance zone) has the following advantages compared with dimensional coordinate tolerancing.

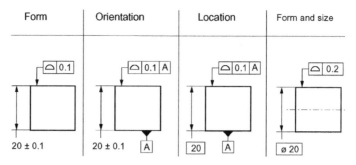

Fig. 4.4 Profile tolerance of surfaces for tolerancing form, orientation, location or form and size

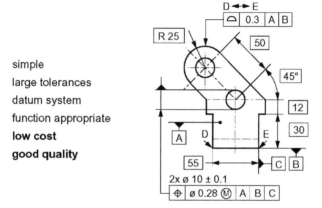

simple

large tolerances

datum system

function appropriate

low cost

good quality

Fig. 4.5 Profile tolerancing

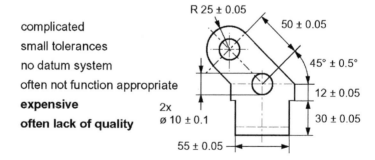

complicated

small tolerances

no datum system

often not function appropriate

expensive

often lack of quality

Fig. 4.6 Dimensional coordinate tolerancing ($\pm$ tolerancing)

Relationships to one or more datum(s) can be indicated unequivocally. Functional relationships can be better (directly) indicated. Therefore function-related tolerancing with the largest possible tolerances is possible (see 20.6.4), production costs are lower and product quality is higher. See also 6.8.

5

Tolerancing of Cones

5.1 General

With cones, the following characteristics may be toleranced:

- surface profile (form) of the cone (conicity);
- orientation and radial location of the cone (cone axis) relative to a datum;
- axial location of the cone relative to a datum;
- distance of an end face of the truncated cone relative to another face.

The first three characteristics (deviations) may be toleranced either together (in common) by the profile tolerance of the cone (t in Fig. 5.3), or separately by different tolerances (Figs 5.4 to 5.10).

The cone angle may be specified as the theoretically exact angular dimension or as the theoretically exact cone ratio (Figs 5.1 and 5.2).

In order to define the axial location of the cone, the following may be specified:

- theoretical exact cone diameter (inside the cone) and theoretical exact distance of this diameter from a datum (Fig. 5.3) when the axial deviation is limited by the profile tolerance of the cone, T_x in Fig. 5.3;
- theoretical exact cone diameter (inside the cone) and theoretical exact distance of this diameter from a datum together with a surface profile tolerance of the cone related to this datum and in addition to the (smaller) unrelated profile tolerance of the cone (Fig. 5.4);
- theoretical exact cone end face diameter, when there is a cone end face, and theoretical exact distance of another face from this cone end face datum together with a profile tolerance of the other face (Fig. 5.10);
- theoretical exact gauge dimension (maximum material size) with nominal distance from a datum and permissible deviations of this distance (Fig. 5.5).

In order to define the radial location of the cone (cone axis) relative to a datum axis, the following may be specified:

- form tolerance of the cone related to a datum axis, when the radial deviation shall be limited by the profile tolerance of the cone (Fig. 5.3);
- profile tolerance of the cone related to the datum axis and in addition to the (smaller) unrelated profile tolerance of the cone (Fig. 5.6);
- coaxiality tolerance of the cone axis relative to the datum axis, when the radial deviation may be larger than according to the profile tolerance of the cone (Fig. 5.7);

- run-out tolerance of the cone (instead of the coaxiality tolerance) relative to the datum axis, when the radial deviation may be larger than according to the profile tolerance of the cone (the run-out deviation is easier to inspect than the coaxiality deviation and has, together with the smaller profile tolerance of the cone, the same effect) (Fig. 5.8).

5.2 Form tolerance and dimensioning of the cone

For dimensioning and tolerancing the form of cones see Figs 5.1 and 5.2.

The following figures show on the left the drawing indications and on the right the tolerance zones of the cone.

Instead of the theoretical exact maximum cone diameter another theoretical exact cone diameter (inside the cone) may be specified (Figs 5.3 to 5.5 and 5.9).

The theoretical exact maximum cone diameter should only be specified when this surface is the datum for the axial location of the cone. (Otherwise there must be a

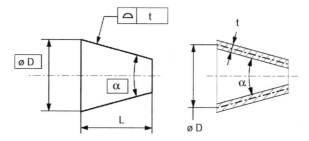

Fig. 5.1 Cone tolerancing by:
profile (form) tolerance *t* of the cone
theoretical exact maximum cone diameter *D*
theoretical exact cone angle α.

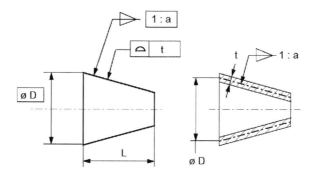

Fig. 5.2 Cone tolerancing by:
profile (form) tolerance *t* of the cone
theoretical exact maximum cone diameter *D*
theoretical exact cone ratio 1:*a*.

distinction between the actual maximum diameter and the theoretical exact maximum diameter. The latter may even not exist on the real workpiece.)

5.3 Tolerancing of the axial location of the cone

Usually cones and truncated cones are not the only features of a workpiece. Then the axial location of a specified diameter of the cone relative to a datum is to be dimensioned and toleranced (Figs 5.3 to 5.5).

According to Fig. 5.3, the axial location of the cone is limited by the profile tolerance of the cone to an axial tolerance T_x.

When the location of the cone may vary more than in Fig. 5.3 larger tolerances may be applied, see T_x in Fig. 5.4.

A frequently used method of tolerancing the axial cone location is shown in Fig. 5.5. It has the advantage of easy inspection when an appropriate gauge can be used. The form of the cone is separately toleranced and to be inspected separately.

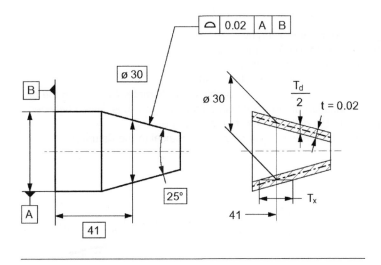

$$T_x = 0.02/\sin 12.5° = t/\sin \alpha/2$$

$$T_d/2 = 0.02/\cos 12.5° = t/\cos \alpha/2$$

Fig. 5.3 Cone tolerancing by:
profile tolerance (0.02) of the cone
theoretical exact cone angle (25°)
theoretical exact cone diameter (ø 30)
theoretical exact distance (41) of this cone diameter (ø 30) from the datum B
the axial location of the cone (ø 30) is limited between $\pm\ T_x/2$ from the distance (41) to the datum B
the axis of the profile tolerance zone is coaxial to the datum A

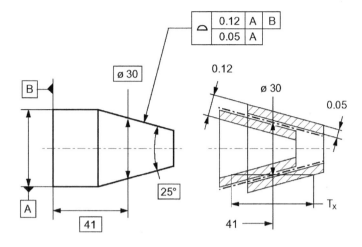

$$T_x = 0.12/\sin 12.5°$$

Fig. 5.4 Cone tolerancing by:
profile tolerance (0.05) of the cone
theoretical exact cone angle (25°)
theoretical exact cone diameter (ø 30)
theoretical exact distance (41) of this cone diameter (ø 30) from the datum A
the axial location of the actual cone diameter (ø 30) is limited between
$\pm T_x/2$ from the distance (41) to the datum B
the axis of the profile tolerance zone is coaxial to the datum A

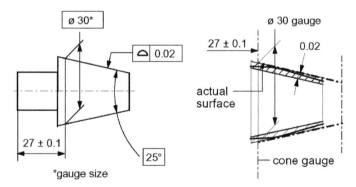

Fig. 5.5 Cone tolerancing by:
profile tolerance (0.02) of the cone
theoretical exact cone angle (25°)
theoretical exact gauge dimension (ø 30)
dimensional tolerance (± 0.1) to limit the actual location of the theoretical
exact gauge dimension (ø 30) from the distance (27) to the end face

5.4 Tolerancing of the orientation and radial location of the cone

According to Fig. 5.3 the orientation and radial location of the cone is limited by the profile tolerance of the cone to a "radial" tolerance $T_d/2$.

When the radial location of the cone may vary more than in Fig. 5.3 a larger profile tolerance (Fig. 5.6) or coaxiality tolerance (Fig. 5.7) or run-out tolerance (Fig. 5.8) may be applied.

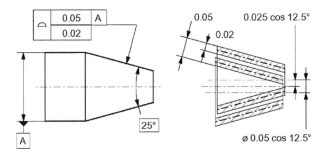

Fig. 5.6 Cone tolerancing by:
profile tolerance (0.02) of the cone
profile tolerance (0.05) of the cone related to a datum axis A
theoretical exact cone angle (25°)
For the sake of clarity the necessary indications defining the axial location of the cone have been omitted in this drawing.

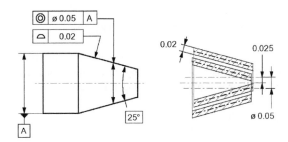

Fig. 5.7 Cone tolerancing by:
profile tolerance (0.02) of the cone
theoretical exact cone angle (25°)
coaxiality tolerance (ø 0.05) for the orientation and radial location of the cone.
For the sake of clarity, the necessary indications defining the axial location of the cone have been omitted in this drawing.
The cone diameter (dimension line) at the leader line arrow of the coaxiality tolerance has no specified dimension in order to indicate that the entire length of the truncated cone is toleranced. The actual axis of the truncated cone must be contained in the coaxiality tolerance zone (ø 0.05).

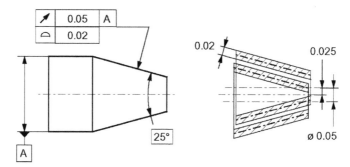

Fig. 5.8 Cone tolerancing by:
profile tolerance (0.02) of the cone
theoretical exact cone angle (25°)
run-out tolerance (0.05) for the orientation and radial location of the cone.
For the sake of clarity, the necessary indications defining the axial location of
the cone have been omitted in this drawing.

5.5 Related profile tolerance for tolerancing form, orientation, radial and axial location of the cone

Figure 5.3 shows tolerancing of a cone when all aspects (form, orientation, radial and axial location of the cone) are limited by the profile tolerance (0.02).

Figures 5.9 and 5.10 show tolerancing of a cone when form, radial and axial location of the cone are limited by different (separate) tolerances.

5.6 Relationship between the cone tolerances

The value of the profile tolerance t of a cone applies in the direction normal to the cone surface (diameter t of imaginary spheres located at the theoretical exact cone, see Fig. 3.2). The profile tolerance can be converted as follows:

- tolerance T_d normal to the cone axis (Fig. 5.11)

$$T_d/2 = t/\cos(\alpha/2), \qquad t = (T_d/2)\cos(\alpha/2);$$

- axial cone tolerance T_x (permissible location of a theoretical exact cone diameter) (Fig. 5.3)

$$T_x = t/\sin(\alpha/2), \qquad t = T_x \sin(\alpha/2);$$

- cone angle α (Fig. 5.11)

$$T_\alpha = \alpha_{max} - \alpha_{min}$$

$$\alpha_{max} = 2\arctan[L\tan(\alpha/2) + T_d/2]/L$$

$$\alpha_{min} = 2\arctan[L\tan(\alpha/2) - T_d/2]/L$$

$$t = L\cos(\alpha/2)\{\tan[(\alpha/2) - (T_\alpha/4)]\}$$

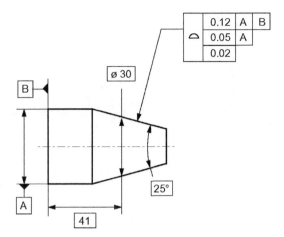

Fig. 5.9 Cone tolerancing by:
profile tolerance (0.02) of the cone limits (form of the cone)
profile tolerance (0.05) of the cone related to the datum axis A (limits orientation and radial location)
profile tolerance (0.12) of the cone related to the datums A and B (limits axial location of the cone)

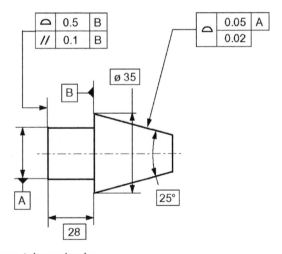

Fig. 5.10 Cone tolerancing by:
profile tolerance (0.02) of the cone (limits form of the cone)
profile tolerance (0.05) of the cone related to the datum axis A (limits orientation and radial location)
profile tolerance (0.5) of another face related to the datum B (limits axial location of the cone)
parallelism tolerance (0.1) of the other face related to the datum B (cone end face) (limits parallelism of the faces)

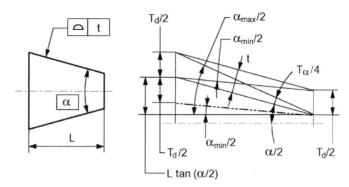

Fig. 5.11 Conversion of cone profile tolerance t and cone angle tolerance $\pm T_\alpha/2$ or cone angle limit values α_{min}, α_{max}

When the cone ratio is 1:3 or smaller, or when the cone angle α is 20° or smaller, the difference between t and $T_d/2$ is less than 2% and therefore practically negligible.

ISO 1947 still uses cone angle tolerances T_α and cone diameter tolerances T_d.

ISO 5166 still uses axial cone tolerances T_x.

These tolerances can be converted into cone profile tolerances t as given above.

6

Positional Tolerancing

6.1 Definition

The positional tolerance is defined in ISO 1101 and positional tolerancing is described in ISO 5458.

In the method of positional tolerancing (for the location of features) theoretical exact dimensions and positional tolerances determine the location of features (points, axes, median faces and plane surfaces) relative to each other or in relation to one or more datum(s). The tolerance zone is symmetrically disposed about the theoretical exact location.

By virtue of this definition positional tolerances do not accumulate where theoretical exact dimensions are arranged in a chain (Figs 6.1, 6.2 and 18.55). This contrasts with dimensional tolerances arranged in a chain.

6.2 Theoretical exact dimensions

According to ISO 1101 theoretical exact dimensions are to be indicated in rectangular frames. However, the following theoretical exact dimensions that determine the theoretical exact locations of positional tolerance zones are not to be indicated:

(a) theoretical exact angles between features (e.g. holes) equally spaced on a complete pitch circle (Figs 6.4 and 6.5); and

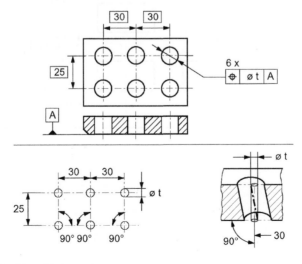

Fig. 6.1 Positional tolerancing

(b) theoretical exact dimensions 90° and 0°, 180° or distance 0 (features located theoretically on a straight line) between
- positional toleranced features not related to a datum (Figs 6.2, 20.19 and 20.70),
- positional toleranced features related to the same datum(s) (Figs 3.24 and 6.4),
- positional toleranced features and their related datum(s) (Fig. 6.1).

When the positional toleranced features are drawn on the same centre line, they are regarded as related features having the same theoretical exact location (Fig. 6.4), unless otherwise specified, e.g. by relation to different datums (Fig. 3.23), or by an appropriate note on the drawing (Fig. 6.5). In Fig. 6.5 the locations of the two patterns of features (holes) are independent from each other. See also 20.6.5.

However, datums are not necessarily perpendicular to each other when they are not in the same datum system. (This applies to all datums of any geometrical tolerancing.)

6.3 Form of the positional tolerance zone

The width of the tolerance zone has the direction of the leader line arrow (which connects the toleranced feature with the tolerance frame) (Fig. 6.2).

With the drawing indication according to Fig. 6.2 without the "axis" indications, the median faces of the two opposite plane surfaces would be toleranced. The tolerance zone would then be the space between two parallel planes of distance 0.3 or 0.1 respectively along the entire width and height of the hole.

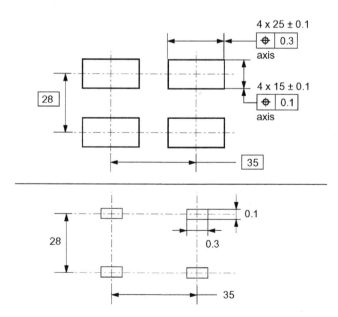

Fig. 6.2 Positional tolerance zones with perpendicular cross-sections

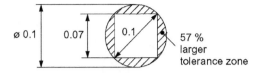

Fig. 6.3 Comparison of tolerance zones with round and square cross-section

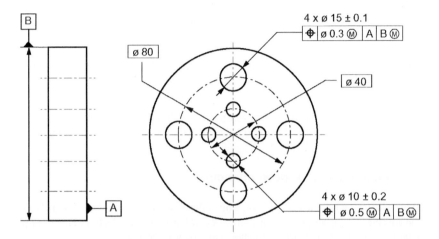

Fig. 6.4 Positional toleranced features (holes) equally spaced on a complete pitch circle, drawn on the same centre line, and therefore regarded as related features having the same theoretical exact location

However, in ISO 5458:1987 for these examples the tolerance zone is referred to as being applicable to the axes of the (rectangular) holes. Therefore, in order to avoid misunderstandings, it is recommended that in the case of rectangular features the indication "axis" should be added to the tolerance frame when the axes are toleranced (Fig. 6.2), or the indication "median faces" when the median faces are toleranced.

When the tolerance value is preceded by the symbol ø the tolerance zone is cylindrical. For cylindrical features of mating parts, the tolerance zone is usually cylindrical because

- the function permits the same deviation in all directions (multidirectional) from the theoretical exact location;
- manufacturing causes multidirectional deviations.

Positional tolerancing provides an indication of cylindrical tolerance zones. This contrasts with dimensional tolerancing, which can only generate rectangular tolerance zones. Cylindrical tolerance zones are 57% larger than the corresponding square tolerance zones (Fig. 6.3).

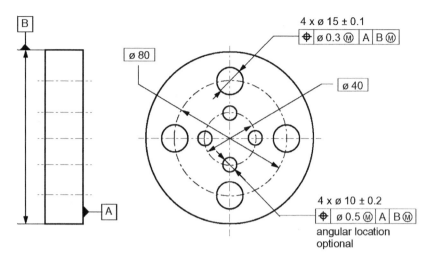

Fig. 6.5 Positional toleranced features (holes) drawn on the same centre line, but with angular location of the two patterns relative to each other being optional

6.4 Positional tolerances on a circle

With tolerancing of hole distances on a pitch diameter positional tolerancing provides the advantage of specifying the theoretical exact location of the feature. This is a prerequisite for proper tolerance calculation and proper inspection.

Dimensional tolerancing, however, results in tolerance accumulation and, on a closed pitch circle, in contradictory tolerances. Therefore proper tolerance calculations and proper inspections are not possible with dimensional tolerancing on a closed pitch circle.

6.5 Positional tolerances related to a datum

Positional tolerances may be related to one or more datum(s) in an unequivocal way (Figs 6.1 and 6.4). Where positional tolerance zones are perpendicular to the related datum, the theoretical exact angle 90° need not to be indicated in the drawing (see 6.2 and Fig. 6.1). Where this relationship is omitted (Fig. 6.1 without datum indication) the positional tolerance zones are parallel to each other, but they do not necessarily need to be perpendicular to the faces.

6.6 Tolerance combinations

6.6.1 Combination of dimensional coordinate tolerancing and positional tolerancing

When features within a group are individually located (relative to each other) by positional tolerancing and their pattern is located by coordinate dimensioning and tolerancing (± limit deviations), each requirement shall be met independently (Fig. 6.6) (ISO 5458).

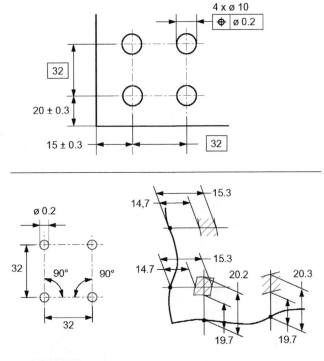

Fig. 6.6 Dimensional coordinate tolerances adjacent to positional tolerances according to ISO 5458

However, the interpretation of the ± limit deviations (±0.3) in Fig. 6.6 is not unique. For example, the distances between the actual axis of each left hole and the left edge can be interpreted as to be between the two limits of size, 19.7 and 20.3.

Another possible interpretation is that the distance between the actual axis and a contacting plane must be between the two limits of size. But still there is an uncertainty as to which contacting plane is primary and which is secondary.

The interpretation of the positional tolerance between the holes is unique. The actual axis of each hole shall be within the cylindrical positional tolerance zone ø 0.1. The positional tolerance zones are located in their theoretical exact location relative to each other.

According to ANSI Y14.5 - 1973 and the former BS 308 this drawing indication has a different meaning. The dimensional coordinate tolerances apply to the location of the theoretical exact pattern of positional tolerances. Therefore, according to these standards, the permissible distance of the actual axis from the edge is half of the positional tolerance larger than according to ISO 5458 (Fig. 6.7). The interpretation of the positional tolerance between the holes is the same as according to ISO, see above.

Because of the risk of misunderstanding, this method (dimensional coordinate tolerances adjacent to positional tolerances) should be avoided, and the method of combination of positional tolerances should be applied (see 6.6.2 and 21.2).

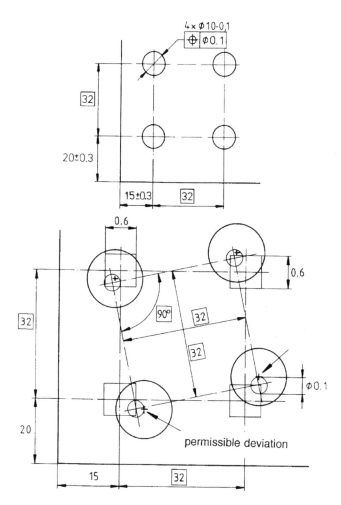

Fig. 6.7 Dimensional coordinate tolerances adjacent to positional tolerances according to ANSI Y14.5 - 1973 and BS 308

6.6.2 Combination of positional tolerances (composite positional tolerancing)

When features within a group are individually located by positional tolerancing and their pattern located by different positional tolerancing, each requirement shall be met independently (Fig. 6.8) (ISO 5458).

For example, in Fig. 6.8, the actual axis of each of the four holes shall be within the cylindrical positional tolerance zone of ø 0.1. The positional tolerance zones are located in their theoretical exact positions in relation to each other and perpendicular to datum A.

The actual axis of each hole shall be within the cylindrical positional tolerance zone of ø 0.6. The positional tolerance zones are located in their theoretical exact positions in relation to the datums A, Y and Z.

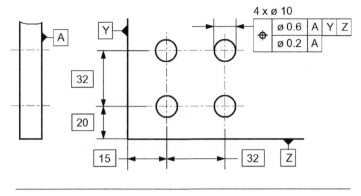

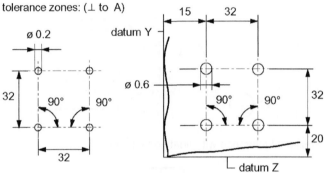

Fig. 6.8 Combination of positional tolerancing with different datums

6.7 Calculation of positional tolerances

A distinction must be made between floating and fixed fasteners (Figs 6.9 and 6.10).
For **floating** fasteners (Fig. 6.9), the positional tolerance t_{cd} of the holes is

$$t_{cd} = M_{ai} - M_{ae} = D_{min} - d_{max}$$

where M_{ai} is the maximum material size of internal dimension, e.g. minimum size of the hole, and M_{ae} is the maximum material size of external dimension, e.g. maximum size of the bolt.

Figure 6.9 shows the extreme locations of the holes of maximum material size that still allow assembly with the bolt of maximum material size. From these locations, the positional tolerance zone ø $(D_{min} - d_{max})$ is derived.

For **fixed** fasteners (Fig. 6.10) the positional tolerance t_{ce} of the holes is

$$t_{ce} = (M_{ai} - M_{ae})/2 = (D_{min} - d_{max})/2$$

Figure 6.10 shows the extreme locations of the holes of maximum material size that still allow assembly with the bolt of maximum material size. From these locations, the positional tolerance zone ø $(D_{min} - d_{max})/2$ is derived.

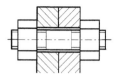

hole ø: max. dim. (D_{max}) limited by pressure
          min. dim. (D_{min}) limited by production method
bolt ø: max. dim. (d_{max}) limited by thread

tolerance for hole position: $D_{min} - d_{max}$

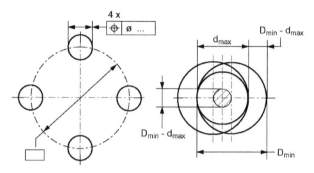

Fig. 6.9 Positional tolerance with floating fastener

hole ø: max. dim. (D_{max}) limited by pressure
          min. dim. (D_{min}) limited by production method
bolt ø: max. dim. (d_{max}) limited by thread

tolerance for hole position: $\dfrac{D_{min} - d_{max}}{2}$

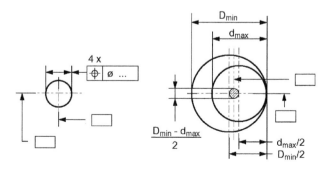

Fig. 6.10 Positional tolerance with fixed fastener

Table 6.1 Positional tolerances for holes according to ISO 273 and screws:

t_{ce} is the positional tolerance for through-hole and threaded hole for stud screw or head screw (Fig. 6.10)

t_{cd} is the positional tolerance for through-hole and threaded bolt with nuts at both ends or head screw with nut (Fig. 6.9)

hole ø D_{min} : ISO 273 medium   tce = $(D_{min} - d) / 2$ tcd = $D_{min} - d$

thread ø	hole ø	tce	tcd	thread ø	hole ø	tce	tcd	thread ø	hole ø	tce	tcd
1	1,2	0,1	0,2	14	15,5	0,75	1,5	64	70	3	6
1,2	1,4	0,1	0,2	16	17,5	0,75	1,5	68	74	3	6
1,4	1,6	0,1	0,2	18	20	1	2	72	78	3	6
1,6	1,8	0,1	0,2	20	22	1	2	76	82	3	6
1,8	2,1	0,15	0,3	22	24	1	2	80	86	3	6
2	2,4	0,2	0,4	24	26	1	2	85	91	3	6
2,5	2,9	0,2	0,4	27	30	1,5	3	90	96	3	6
3	3,4	0,2	0,4	30	33	1,5	3	95	101	3	6
3,5	3,9	0,2	0,4	33	36	1,5	3	100	107	3,5	7
4	4,5	0,25	0,5	36	39	1,5	3	105	112	3,5	7
4,5	5	0,25	0,5	39	42	1,5	3	110	117	3,5	7
5	5,5	0,25	0,5	42	45	1,5	3	115	122	3,5	7
6	6,6	0,3	0,6	45	48	1,5	3	120	127	3,5	7
7	7,6	0,3	0,6	48	52	2	4	125	132	3,5	7
8	9	0,5	1	52	56	2	4	130	137	3,5	7
10	11	0,5	1	56	62	3	6	140	147	3,5	7
12	13,5	0,75	1,5	60	66	3	6	150	158	4	8

positional tolerance t_c:
only through holes $t_c = t_{cd} - t$
through holes and threaded holes $t_c = t_{ce} - t$

ISO 4759-1:
d ≤ 8: t = 0,002 l + 0,05
d > 8: t = 0,0025 l + 0,05

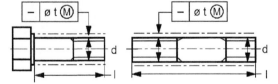

The formulae are also applicable for non-cylindrical features (e.g. keys and key ways) (see also 20.12 and 20.13).

The indicated positional tolerances t_{ce} and t_{cd} in Table 6.1 do not take account of the straightness tolerances of the screw and hole, the minimum clearance between bolt and hole, or any measurement uncertainty. If necessary, the values are to be decreased accordingly. For head screws the under-head fillet may require chamfered holes or washers. Table 6.1 bottom shows the formulae for the straightness tolerances t of the screws according to ISO 4759-1. Accordingly for the positional tolerances t_c the values t_{cd} and t_{ce} are to be decreased by t.

6.8 Advantages of positional tolerancing

Positional tolerancing has the following advantages compared with dimensional coordinate tolerancing:

- Functional relationships are better (directly) indicated. Relationships to one or more datum(s) can be indicated unequivocally. Therefore function-related tolerancing with the largest possible tolerances is possible (see 20.6.4).
- There is the possibility of indicating cylindrical tolerances. Thereby 57% larger tolerances are possible compared with tolerances of rectangular cross-section (Fig. 6.3). In many cases this corresponds to the function-related tolerancing, e.g. mating of cylindrical bolts with cylindrical holes.
- There is a simple application of the maximum material requirement with additional gain of tolerance.
- There is the possibility of application of the projected tolerance zone method (see 7). Thereby function-related tolerancing (with the largest possible tolerances) becomes possible.
- There is no tolerance accumulation when theoretical exact dimensions are arranged in a chain. This enables simple tolerance calculations directly according to the function. In some cases, e.g. holes on a pitch diameter, in practice only positional tolerances allow proper tolerance calculations, for function-related, manufacturing-related and inspection-related tolerancing (see 20.6.4).

7

Projected Tolerance Zone

Positional tolerances of threaded holes and through-holes for fixed fasteners (Figs 6.10 and 7.1) are usually calculated from the maximum diameter of the bolt and the minimum diameter of the through-hole (see 6.7).

A prerequisite for these calculations is that the axis is perpendicular to the joint surface. In the extreme case (Fig. 7.2) the parts still fit.

When the axis of the threaded hole deviates from the perpendicular orientation (but remains in the positional tolerance zone) the parts do not fit (Fig. 7.3).

However, when the positional tolerance zone is located outside the part and applies to the external projection of the feature (axis) (Fig. 7.4), the part fits in any case (Fig. 7.5). The external projection has the length of the through-hole in the case of head screws (Fig. 7.6), and the length of the outstanding part of the stud or dowel pin (Fig. 7.7). The projected tolerance zone applies only to these lengths, not to the (length of the) axis within the workpiece.

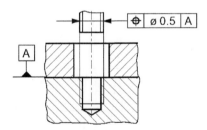

Fig. 7.1 Positional tolerance of a threaded hole

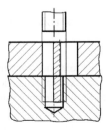

Fig. 7.2 Positional tolerance zones of threaded hole and through-hole

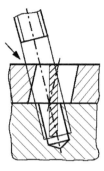

Fig. 7.3 Orientation deviations taking advantage of the positional tolerance zones

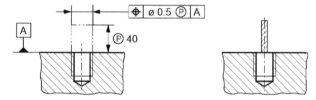

Fig. 7.4 Projected tolerance zone

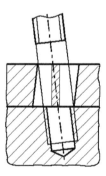

Fig. 7.5 Taking advantage of the positional zone by inclined axes

ISO 10 578 defines the projected tolerance zone as follows. The **projected (positional) tolerance zone** applies to the external projection of the feature indicated on the drawing by the symbol Ⓟ placed in the tolerance frame after the positional tolerance of the toleranced feature. The minimum extent and the location of the projected tolerance zone are shown in the corresponding drawing view by Ⓟ preceding the projected dimension (Fig. 7.4).

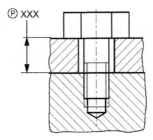

Fig. 7.6 Projected tolerance zone for threaded hole and head screw

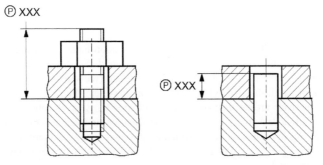

Fig. 7.7 Projected tolerance zone for threaded hole and stud or for hole and pin

According to ASME Y14.5 the length of the projected tolerance zone may be indicated in the tolerance frame after the symbol Ⓟ, e.g.:

| ⊕ | ⌀ 0.5 Ⓟ 40 | A |

The exact definition of the axis to be extended (projected) thus far is not standardized (see 3.5). From a practical point of view, the following definitions of the axis may be applicable:

- for plain holes (e.g. interference fits): the axis of the contacting cylinder;
- for threaded holes: the axis of a (nearly) geometrical ideal screw of maximum material sizes (go gauge).

8

Substitute Elements

8.1 General

Substitute elements (substitute features) are imaginary geometrical ideal features (e.g. straight line, circle, plane, cylinder, sphere, cone and torus). Their location, orientation and (if applicable) size are calculated from the assessed points of the workpiece surface.

In general, substitute elements are assessed by coordinate measuring machines or form measuring machines. According to the planned ISO Standard on VD&T, the assessment is performed in a right-handed Cartesian (rectangular) coordinate system (Fig. 8.1), or a right-handed cylindrical coordinate system (Fig. 8.2).

Points in space are defined by their coordinates x_0, y_0, z_0 (location vector P) and stored for further data processing (Fig. 8.3). Orientations in space are defined by the components E_x, E_y, E_z of the unit vector (orientation vector N or E). It therefore follows that $N = \sqrt{(E_x^2 + E_y^2 + E_z^2)} = 1$ (Figs 8.3 and 8.4). The orientation vector is always directed out of the material.

Table 8.1 shows the vectors needed to define the substitute element. The indicated minimum numbers of points to be assessed apply to geometrical ideal features. As the actual features always exhibit form deviations, many more points are necessary for measuring

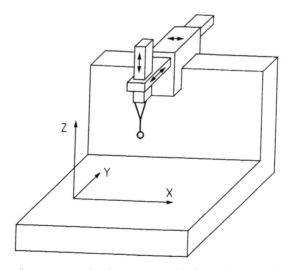

Fig. 8.1 Coordinate measuring instrument with Cartesian coordinate system

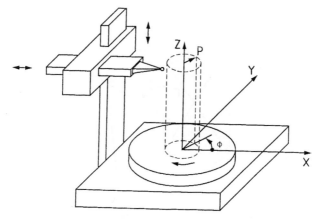

Fig. 8.2 Coordinate measuring instrument with cylindrical coordinate system

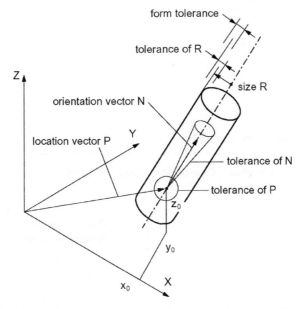

Fig. 8.3 Definition of a substitute cylinder (location vector, orientation vector and substitute size)

an actual feature of the workpiece. The workpiece is sensed at a number of points and, after tracer radius correction, the calculated coordinates of the actual feature points are stored. From the stored coordinates, the substitute element is calculated (Fig. 8.5).

The calculation of the substitute element may be done according to the following procedure:

(a) Gauss (the sum of squares of deviations of the sensed points from the substitute element is minimized);

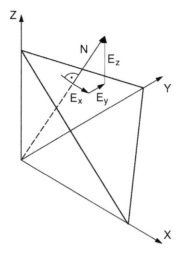

Fig. 8.4 Orientation vector of a plane

Table 8.1 Substitute elements

Element	Location X_0, Y_0, Z_0	Orientation E_x, E_y, E_z	Size	Remarks	Minimum number of points
Plane	X	X		Point on plane Normal to plane	3
Sphere	X		X	Centre point Diameter	4
Cylinder	X	X	X	Point on axis Orientation of axis Diameter	5
Cone	X	X	X	Point on axis Orientation of axis Cone angle	6
Torus	X	X	XX	Centre point Orientation of axis Diameters of pipe and ring	7

Table 8.1 (Cont.)

Element	Location X_0, Y_0, Z_0	Orientation E_x, E_y, E_z	Size	Remarks	Minimum number of points
Straight line	X			Point on straight line	2
		X		Orientation of straight line	
Circle	X			Centre point	3
		X		Normal to plane	
			X	Diameter	

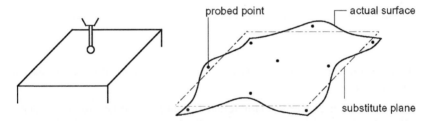

Fig. 8.5 Sensed points and substitute element of a plane workpiece surface

(b) Chebyshev (the maximum deviation of the sensed points from the substitute element is minimized);
(c) Contacting element (only for bounded geometrical elements, maximum inscribed geometrical element for holes or minimum circumscribed geometrical element for shafts).

These methods may yield different results, although they use the same sensed points. Only the Gaussian method is always unique. The planned ISO Standard will probably require that, unless otherwise specified, the Gaussian method is to be applied.

8.2 Vectorial dimensioning and tolerancing

Substitute elements are defined by vectors (location vector, orientation vector and size(s)) in a substitute datum system. Dimensioning and tolerancing with the aid of substitute elements is called vectorial dimensioning and tolerancing (VD&T).

For the use of VD&T the following definitions and drawing indications must be agreed upon. As there is not up to this time a standard on this subject, they must be specified in a company standard or by reference to this book.

Fig. 8.6 Identification of a feature (No. 2) on the drawing

Fig. 8.7 Drawing indication of a substitute intersection point (of substitute elements Nos 2 and 3)

Fig. 8.8 Drawing indication of specified points (of feature Nos 5 and 6)

Actual feature: Feature (actual surface, actual axis or actual median face) obtained by measurement.

Nominal feature: Geometrical ideal feature (nominal surface, nominal axis or nominal median face) as defined by the drawing. To identify the feature on the drawing an enumeration with symbolization according to Fig. 8.6 may be used.

Substitute element: Geometrical ideal element of similar form to the nominal feature (e.g. plane, sphere, cylinder, cone or torus) and derived from the measured data using the Gaussian algorithm.

Substitute axis: Axis of a substitute cylinder, cone or torus.

Substitute median plane: Median plane between the substitute planes of two opposite parallel plane surfaces.

Substitute intersection point: Point of intersection of a substitute axis and a substitute plane. On drawings, the substitute intersection point may be symbolized by ⊠ followed by the feature numbers combined by the symbol + (Fig. 8.7).

Specified point of a substitute element: Point of a substitute element specified to locate the substitute element relative to the substitute datum system. On drawings the specified point may be symbolized by ⊠ followed by the feature(s) number(s). If the same symbol is applied to more than one feature the feature numbers are separated by commas (Fig. 8.8).

Substitute datum system: Coordinate system of the workpiece derived from substitute elements. Location and orientation of the substitute elements are related to this coordinate system. It is defined by the primary, secondary and tertiary substitute datums as follows.

Primary substitute datum (xy plane) defined by:

- specified substitute plane, or
- plane containing the substitute axis of the primary datum feature, the specified substitute intersection point (apart from the substitute axis, intersection of another substitute axis with a substitute plane, see Fig. 8.17).

Secondary substitute datum (xz plane) defined by a plane perpendicular to the primary datum and containing:

- the intersection line of the substitute plane of the secondary datum feature with the primary datum, or
- the substitute axis of the primary datum feature (which thereby also becomes a secondary datum feature, see Fig. 8.17), or
- the line connecting two substitute intersection points specified as secondary datums, this line being not normal to the primary datum.

Tertiary substitute datum (yz plane) defined by a plane perpendicular to the primary and to the secondary datums, and containing:

- the point of intersection of the substitute plane of the tertiary datum feature with the primary and secondary datum, or
- the substitute intersection point specified as tertiary datum.

To identify a substitute datum for reference purposes, a capital letter preceded by S for substitute is enclosed in a rectangular frame connected to a datum triangle (Fig. 8.9).

The substitute datum system is indicated by three subsequent rectangular compartments (Fig. 8.10). The sequence from left to right indicates the order of primary, secondary and tertiary datum.

If a datum is defined by a combination of an axis and a substitute intersection point or by two substitute intersection points, the datum letters in the datum system frame are combined by a hyphen (Fig. 8.11).

Fig. 8.9 Drawing indication of a substitute datum

| SA | SB | SC |

Fig. 8.10 Drawing indication of a substitute datum system

| SA - SB | |

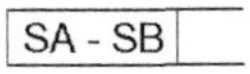

Fig. 8.11 Drawing indication of a common substitute datum

The mathematically positive coordinate system according to ISO 841 applies (Fig. 8.12).

Linear substitute size: These are the diameters of substitute spheres, cylinders, cones and tori.

The *linear substitute size deviation* is the difference between the actual substitute size and the nominal substitute size.

The *linear substitute size tolerance* is the difference between the upper and lower limit of the linear substitute size deviations.

Angular substitute size: These are the cone angles (apexes) of substitute cones.

The *angular substitute size deviation* is the difference between the actual substitute cone angle and the nominal substitute cone angle.

The *angular substitute size tolerance* is the difference between the upper and lower limit of the angular substitute size deviations.

Substitute location: This is defined by the *substitute location vector*. It indicates the location of a *specified point* of the substitute element in the substitute datum system. The specified point can be:

- substitute intersection point;
- point of the substitute element specified to lie:
  - in the case of a plane, on a straight line defined by the two coordinates and perpendicular to the plane that is defined by the two pertinent coordinate axes (Fig. 8.13);
  - in the case of an axis, on a plane defined by one coordinate and parallel to the plane defined by the two other coordinate axes (Fig. 8.14);

Fig. 8.12 Mathematically positive coordinate system

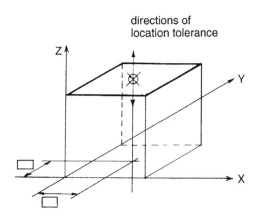

Fig. 8.13 Substitute location of a substitute plane

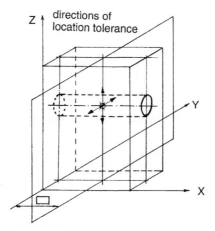

Fig. 8.14 Substitute location of a substitute cylinder

• centre of a substitute sphere;
• centre of a substitute torus.

The *substitute location deviation* is the difference between the actual substitute location vector and the nominal substitute location vector. It can be expressed by its three components in the directions of the axes of the substitute datum system.

The *substitute location tolerance* is the difference between the extreme permissible substitute location deviations expressed in the three directions of the axes of the substitute datum system. In the case of a specified point of a substitute plane two coordinates are theoretically exact, and only one is to be toleranced (Fig. 8.13). In the case of a specified point of a substitute axis one coordinate is theoretically exact and two are to be toleranced (Fig. 8.14).

Substitute orientation: This is defined by the substitute orientation vector. It indicates the orientation of the substitute plane or the substitute axis within the substitute datum system. The substitute orientation vector is normal to the substitute plane or parallel to the substitute axis (cylinder, cone or torus) and points

• in the case of a plane away from the material;
• in the case of a cone in the direction of the greater size.

The *substitute orientation vector* E has a magnitude of 1 and is defined by its components E_x, E_y, E_z, which are in the directions of the axes of the substitute coordinate system (substitute datum system).

The *substitute orientation deviation* is the difference between the actual substitute orientation vector and the nominal substitute orientation vector. It can be expressed by its three components in the directions of the axes of the substitute datum system.

The *substitute orientation tolerance* is the difference between the extreme permissible substitute orientation deviations expressed in the directions of the axes of the substitute datum system. In order to avoid overspecification only two of the three components of the substitute orientation vectors can be toleranced (the third is already determined by the difference from 1).

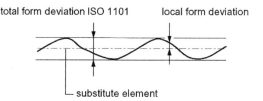

Fig. 8.15 Local form deviation and total form deviation

Form: The nominal form is defined by the drawing or by indication in a table. In case of a bounded form (sphere, cylinder, cone or torus) and indication in a table it should be stated whether it is an internal form (hole) H or an external form (shaft) S.

The *local form deviation* is the distance of the workpiece feature from the substitute element at the considered location. The *total form deviation* is the range of the local form deviations (see Fig. 8.15).

The *form tolerance* is the value of the permitted total form deviation.

Note: According to ISO 1101 the form tolerance zone has the direction of the minimum zone. With vectorial tolerancing the form tolerance zone has the direction of the substitute element. There may be small differences in the measurement results between the two methods.

Surface roughness: The same applies as with conventional dimensioning and tolerancing. See ISO 4287 and ISO 4288.

VD&T requires the following specifications:

- substitute datum system (substitute elements, substitute intersection points);
- location and locational tolerance of one point (specified point) of the substitute element (Table 8.1);
- orientation and orientational tolerance of the substitute element (Table 8.1);
- size(s) and size tolerance(s) of the substitute element (Table 8.1);
- cone angle and cone angle tolerance of the substitute cone (Table 8.1);
- form tolerance (superimposed on the substitute element).

These specifications may be listed according to their feature numbers (Fig. 8.16).

Figure 8.17 shows an example of VD&T. The workpiece coordinate system is determined by:

- axis of substitute feature No. 1, substitute intersection point of axis of substitute feature No. 2 and substitute feature No. 3 (SA-SB) to establish the *xy* plane;
- axis of substitute feature No. 1 (SA), perpendicular to the *xy* plane to establish the *xz* plane;
- substitute intersection point of the axis of substitute feature No. 2 and substitute feature No. 3 (SB), perpendicular to the *xy* and *xz* planes to establish the *yz* plane.

With the enumeration of specified points the sign "+", e.g. "2 + 3", indicates that it is the substitute intersection point of two substitute elements, the sign ",", e.g. "5, 6", indicates that there are two different specified points of the same nominal location.

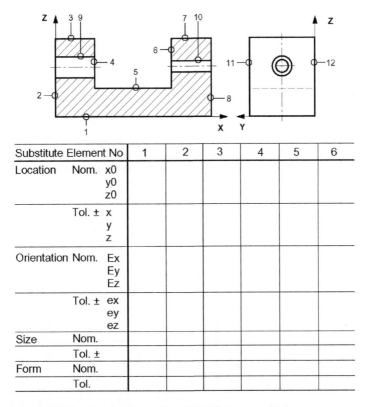

Substitute Element No		1	2	3	4	5	6
Location	Nom. x0						
	y0						
	z0						
	Tol. ± x						
	y						
	z						
Orientation	Nom. Ex						
	Ey						
	Ez						
	Tol. ± ex						
	ey						
	ez						
Size	Nom.						
	Tol. ±						
Form	Nom.						
	Tol.						

Fig. 8.16 Principle of a drawing with vectorial sizes and tolerances

The location tolerances apply to the specified points of the substitute elements, e.g. for the substitute element No. 2 determined by the specified point $2 + 3$ in the y direction (because in the z direction it coincides by definition of the workpiece coordinate system with the coordinate $z = 0$) and for substitute element No. 5, determined by the specified point 5 in the y and z direction (in the x direction the point is specified by the nominal value 45 which determines the point of the substitute element axis at which the location deviation shall be assessed).

The orientation tolerances apply to the substitute elements themselves (not to specified points). The axis of substitute element No. 1 has no tolerance, because it establishes the x direction of the workpiece coordinate system.

Regarding orientation, the permissible deviations are only specified for two components of the orientation vector, because the permissible deviation of the third component is already determined by the condition that the orientation vector (resultant of the three components) has the value 1.

When the nominal orientation vector is not parallel to any axis of the coordinate system, permissible deviations may be specified for two components only or for all three components. When permissible deviations are specified for all three components, the extreme deviations cannot occur on the same workpiece, because the resultant of the three actual components has the value 1 (by definition).

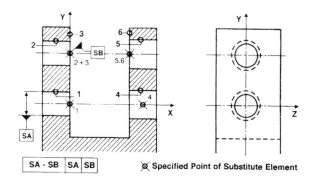

SA - SB	SA	SB	⊠ Specified Point of Substitute Element

Primary Datum	1 - 2 + 3	(Substitute El. and Substitute Intersection Point)
Secondary Datum	1	(Substitute Element)
Tertiary Datum	2 + 3	(Substitute Intersection Point)

Substitute Element No			1	2	3	4	5	6
Location	Nom.	Xo	0	0	0	55	45	45
		Yo	0	35	35	0	35	35
		Zo	0	0	0	0	0	0
	Perm. Dev. ±	x	–	–	–	–	–	0.1
		y	–	0.1	–	0.1	0.1	–
		z	–	–	–	0.1	0.1	–
Orientation	Nom.	Ex	–1	–1	+1	+1	+1	–1
		Ey	0	0	0	0	0	0
		Ez	0	0	0	0	0	0
	Perm. Dev. ±	ex	–	–	–	–	–	–
	x 10^{-3}	ey	–	0.2	0.2	0.2	0.2	0.2
		ez	–	0.2	0.2	0.2	0.2	0.2
Size	Nom.		15	15		20	20	
	Perm. Dev. ±		0.05	0.05		0.05	0.05	
Form	Nom.		Cyl. H	Cyl. H	Plane	Cyl. H	Cyl. H	Plane
	Tol.		0.005	0.005	0.01	0.005	0.005	0.001
Waviness	W$_t$ μm		–	–	–	–	–	–
Roughness	R$_z$ μm		6	6	16	6	6	16
Further Requirements			X	X		X	X	

No	Further Requirements
1	MMVC DIA 14.94 rel. to No 4 MMVC DIA 19.94
2	MMVC DIA 14.94 rel. to No 5 MMVC DIA 19.94
4	See No 1
5	See No 2

Fig. 8.17 Example of vectorial dimensioning and tolerancing

With form the indication "Cyl H" indicates a cylindrical hole, while "Cyl S" indicates a cylindrical shaft.

In the case of a cylindrical shape it might be useful for process control to distinguish between "conical form deviation" and other (e.g. irregular) form deviations. Then, instead of a cylinder, the form should be specified as a cone of cone angle 0° (and the substitute diameter at half of the feature's length). The deviations will then be separated for conicity (in grade) and other form deviations (in mm).

Further requirements may have to be specified, e.g. the envelope requirement Ⓔ according to ISO 8015, the maximum material requirement Ⓜ or the least material requirement Ⓛ according to ISO 2692. In these cases the geometrical ideal boundary of maximum material virtual size or of least material virtual size may be stated using the following symbols:

DIA   diameter;
DIS   distance of two opposite parallel planes;
E   boundary for the envelope requirement (virtual condition) (see 10);
M   boundary for the maximum material requirement (maximum material virtual condition) (see 9);
L   boundary for the least material requirement (least material virtual condition) (see 11).

The theoretical exact dimensions to determine the locations of the virtual conditions (in cases of Ⓜ or Ⓛ) are given by, or are to be derived from, the nominal location vectors of the substitute elements. The orientations of the virtual conditions are given by the nominal orientation vectors.

Example 1

Indication: No. 5 E DIA 19.9
Meaning: The actual feature No. 5 shall not violate the geometrical ideal boundary (cylinder) of ø 19.9.

Example 2

Indication: No. 1 M DIA 14.94 rel. to No. 4 M DIA 19.94
Meaning: The actual features No. 1 and No. 4 shall not violate the geometrically ideal boundaries (cylinders, maximum material virtual condition) of ø 14.94. The location and orientation of the boundaries are theoretically exact (coaxial) as given by the location and orientation vectors (see Fig. 8.17).

Example 3

Indication: No. 8 – No. 9 M DIS 20 rel. to No.10 – No. 11 M DIS 25
               No. 9 see No. 8
               No. 10 see No. 8
               No. 11 see No. 8
Meaning: The actual features Nos 8, 9 and Nos 10, 11 shall not violate the geometrically ideal boundaries (two opposite parallel planes, maximum material virtual condition) of distances 20 and 25 respectively. The location and orientation of the boundaries are theoretically exact

(coplanar median planes) and to be derived from the location and orientation vectors.

Further requirements may concern surface roughness, surface treatment conditions, etc.

It is also possible to use vectorial tolerancing (of substitute elements) for certain selected features while sizes and tolerances according to ISO 8015, ISO 286, ISO 1947 and ISO 1101 (conventional dimensioning and tolerancing) are used for the other features (Fig. 8.18).

The symbols are to be used in a similar way as the symbols according to ISO 1101.

However, in order to avoid misunderstandings, clear distinctions must exist in the drawing specifications regarding sizes, distances, locations, orientations of substitute elements and their tolerances on the one hand, and sizes, distances and their tolerances and orientational and locational tolerances according to ISO 8015, ISO 286, ISO 1947 and ISO 1101 on the other hand. Such distinctions in drawing specifications are at this time not standardized. For the time being, they must be specified, e.g. by a company standard or by reference to this book. Figure 8.18 gives an example of an application. The symbolization used is similar to ISO 1101.

As form deviations are eliminated by calculation from the substitute elements, this method provides a clear distinction between substitute size, substitute location, substitute orientation and form. The deviations are complementary (Figs 8.19 and 8.20, compare with Fig. 3.12). It further provides the direction of the orientation and location

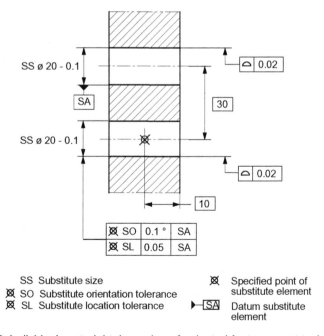

Fig. 8.18 Individual vectorial tolerancing of selected features next to (conventional) tolerancing according to ISO 8015, ISO 286 and ISO 1101 in the same drawing

deviation, and therefore gives the appropriate information for directing of the correction of the production equipment (machine tool).

Therefore vectorial dimensioning and tolerancing may be advantageous for specifying some functional requirements, such as for gears, gear boxes, hydrodynamic bearings and air lubricated bearings, and may be necessary for manufacturing process control (analysing or controlling the manufacturing process).

As deviations in size, orientation and location according to ISO 8015, ISO 1947 and ISO 1101 always include form deviations, they may be less suitable for some of these purposes.

The differences in the measurement results between the two types of sizes and orientation and location deviations from the same workpiece have been found to be up to 100%.

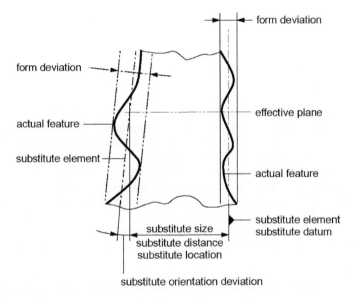

Fig. 8.19 Two nominally parallel surfaces: determination of substitute size, form, substitute orientation and substitute location separately from each other (by using substitute elements)

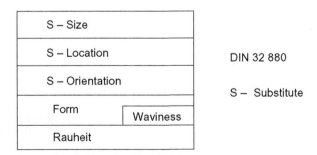

Fig. 8.20 Types of deviations and tolerances of substitute elements (cf. Fig. 3.12)

Which method of dimensioning and tolerancing should be used depends on the functional requirements and on the requirements of the manufacturing process control.

8.3 Comparison of the systems

According to the planned ISO Standard on VD&T and the Draft German Standard DIN 32 880-1, the sizes, distances and deviations of orientation and location relate to the substitute elements and not to the actual features exhibiting form deviations. Therefore they are designated as substitute sizes, substitute distances and substitute deviations (vectorial system). They are to be distinguished from sizes, distances and deviations according to ISO 8015, ISO 286, ISO 1947 and ISO 1101 (conventional system). In the following the differences are described.

8.3.1 Sizes

Figure 8.21 shows a cylinder with form deviations. The actual local sizes according to ISO 8015 and ISO 268 vary between the minimum and maximum actual local sizes. The actual substitute size in this case is the diameter of the substitute cylinder.

Figure 8.22 shows a cone with form deviations. The actual local cone angles according to ISO 8015 and ISO 1947 vary between the maximum and minimum actual local cone angles. The actual substitute cone angle is the cone angle of the substitute cone (which deviates from the nominal cone angle).

Figure 8.23 shows a bar with holes. The bar has deviations from straightness. The actual local distances between the edge and the hole centres according to ISO 8015 are approximately the same at all locations. The distances of the hole centres from the substitute plane, however, differ throughout.

Depending on the data processing program incorporated, the coordinate measuring machine can also assess actual local sizes (length sizes and angular sizes) according to ISO 8015.

However, in this case, difficulties may arise from the fact that an unambiguous standardized definition of the actual local size did not hitherto exist. Therefore the coordinate

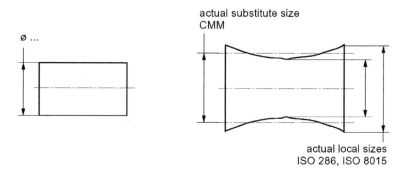

Fig. 8.21 Cylinder with form deviations, actual local sizes and actual substitute size

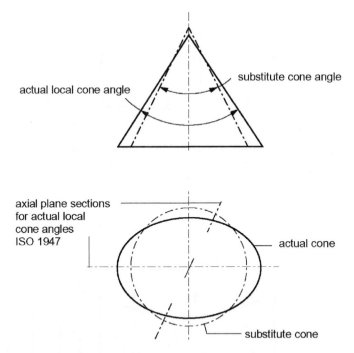

Fig. 8.22 Cone with form deviations, actual local cone angles and actual substitute cone angle

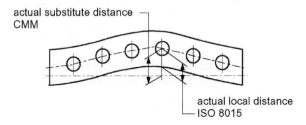

Fig. 8.23 Bar with holes and deviations from straightness, actual local distances and distances from the substitute plane

measuring machines may still use different definitions of actual sizes. The distance between two parallel planes, for example, is assessed as:

(a) distance perpendicular to a datum substitute plane to be chosen;
(b) distance between the corners of the substitute planes;
(c) distance between the centre of gravity S of one substitute plane and the other substitute plane;
(d) distance between two parallel substitute planes (Fig. 8.24).

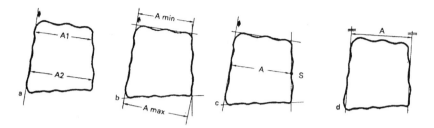

Fig. 8.24 Coordinate measuring technique, former procedures to assess the distance between two parallel surfaces (according to Wirtz)

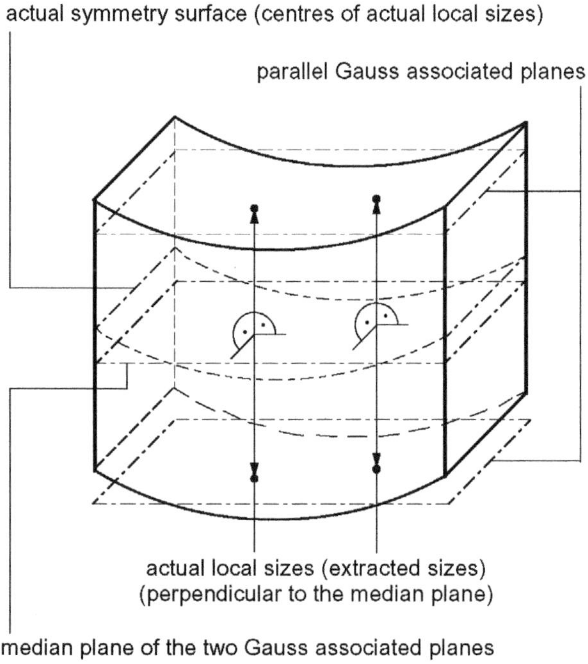

actual symmetry surface (centres of actual local sizes)

parallel Gauss associated planes

actual local sizes (extracted sizes)
(perpendicular to the median plane)

median plane of the two Gauss associated planes

Figure 8.25 Measurement of thickness; actual local size

The procedures for assessing actual local sizes that are now standardized in ISO 14 660-2 are the following:

(a) *Distance between two parallel faces, e.g. wall thicknesses:* The actual local size is the distance of two opposite points, the connection of the two points being perpendicular to the substitute median plane. The substitute median plane is the median plane between the parallel substitute planes of the two opposite surfaces (Fig. 8.25).

(b) *Diameter:* The actual local diameter (size) is the distance of two opposite points estimated in a section perpendicular to the substitute cylinder axis, the line connecting

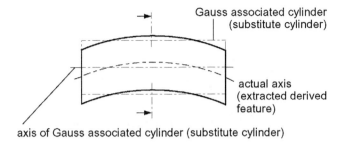

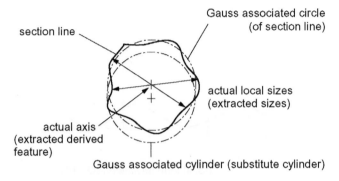

Fig. 8.26 Measurement of actual local diameter

the two points meeting the centre of the Gaussian regression (least squares) circle in this plane (Fig. 8.26).

It must be noted, however, that there are also other possible definitions to agree upon. Opinions still differ as to which substitute circle should be used. The Gaussian regression circle has the advantage of needing the least number of traced points and always being unique. The Chebyshev substitute circle has the advantage of being standardized in ISO 1101 for the assessment of roundness but the disadvantage of needing a much larger number of traced points and not always being unique. The contacting substitute circle (maximum inscribed or minimum circumscribed) has the advantage of being in conformance with ISO 5459 for the definition of datums, but has the disadvantage of not always being unique. The author recommends the use of the least squares method as standardized in ISO 14 660-2 (see 3.5).

8.3.2 Form deviations

When the Chebyshev criterion (see 3.5 and 8.1) is used for obtaining a substitute element and the form deviation is estimated in relation to this substitute element, the approach is the same as according to ISO 1101.

When other criteria are used, the value of the form deviation obtained may be larger.

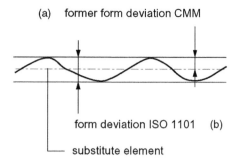

Fig. 8.27 Form deviation: (a) former coordinate measurement; (b) according to ISO 1101

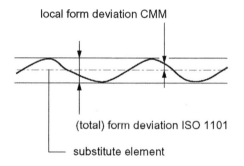

Fig. 8.28 Local form deviation

Old coordinate measuring machines sometimes measure the form deviation as the maximum distance of the traced points from the substitute element (Fig. 8.27). This is then only half the value of the form deviation according to ISO 1101.

The distance of the traced point from the substitute element is the actual local form deviation (Fig. 8.28). The form deviation according to ISO 1101 is the range of the actual local form deviations.

8.3.3 Deviations of orientation and location

With coordinate measuring machines, deviations of orientation are normally obtained as an inclination of the substitute element, sometimes expressed as a distance (Fig. 8.29). In this procedure the form deviation is eliminated. The sign of the orientation deviation indicates the direction of the orientation deviation. These are the main differences from the deviation of orientation according to ISO 1101.

Figure 8.30 shows a workpiece with deviations of coaxiality. The deviation of coaxiality according to ISO 1101 also includes the form deviation of the axis of the cylinder to be measured, and is therefore greater than the coaxiality deviation of the substitute element.

Figure 8.31 shows a workpiece with a coaxiality deviation according to ISO 1101 and practically no coaxiality deviation of the substitute element.

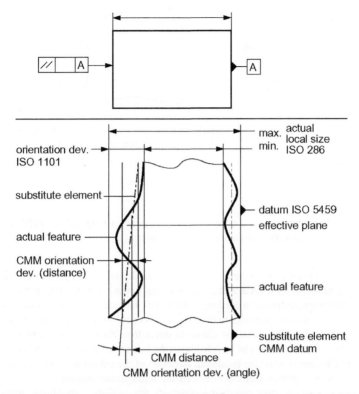

Fig. 8.29 Deviation of orientation (parallelism) according to ISO 1101, and orientational deviation of the substitute element

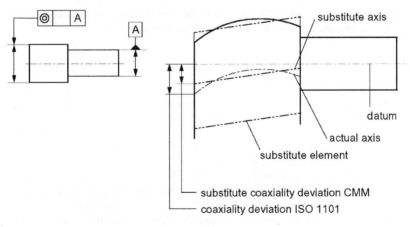

Fig. 8.30 Coaxiality deviation according to ISO 1101, and coaxiality deviation of the substitute element

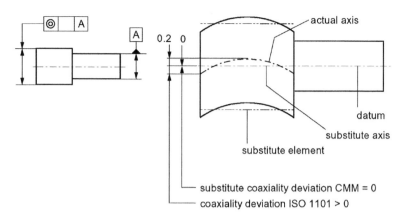

Fig. 8.31 Workpiece with a coaxiality deviation according to ISO 1101 but with practically no coaxiality deviation of the substitute element

Figure 8.32 shows a workpiece with a positional deviation. The positional deviation according to ISO 1101 is greater than the positional deviation of the substitute elements.

The positional deviation according to ISO 1101 includes the form deviation of the toleranced feature, and is related to the datum that contacts the datum feature. The positional deviations of the substitute elements differ from those according to ISO 1101 because the form deviations of both features (the toleranced feature and the datum feature) are eliminated when the positional deviation of the substitute element is assessed and the substitute datum element intersects the actual datum feature.

When appropriate data processing programs are available, coordinate measuring machines can also calculate deviations of orientation or location according to ISO 1101.

However, it is therefore necessary to specify the definition of the actual axis of a cylindrical or conical feature and of the actual median face of a feature of size composed of two parallel opposite surfaces. According to ISO 1101, it is assumed that the actual axis and the actual median face follow the form of the feature and therefore deviate from a straight line or plane (see 3.5 and Figs 8.25 and 8.26).

8.3.4 Datum systems

According to the planned ISO standard on VD&T, the datum systems for workpieces for the determination of location and orientation of the substitute elements (substitute datum system) may be defined as follows (see also 8.2: substitute datum system):

Primary datum: Substitute plane of the primary datum feature (xy plane, if not otherwise specified);

Secondary datum: Plane normal to the primary datum containing the line of intersection of the primary datum and the substitute plane of the secondary datum feature (xz plane, if not otherwise specified);

Tertiary datum: Plane normal to the primary datum and to the secondary datum containing the point of intersection of all three substitute planes (yz plane, if not otherwise specified).

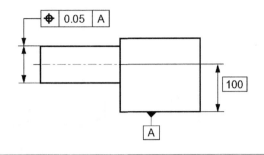

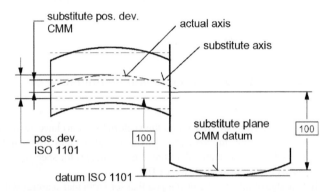

Fig. 8.32 Positional deviation according to ISO 1101, and positional deviation of the substitute element

This datum system for substitute elements is based on substitute planes and their intersections. Workpiece datum systems according to ISO 1101 and ISO 5459, however, are based on contacting elements contacting the highest points of the workpiece surfaces and directed according to the minimum rock requirement.

In order to obtain comparable results using coordinate measuring machines and measuring deviations according to ISO 1101 the substitute datum system must be transformed.

Two options are available for this transformation. One involves moving the three coordinate system planes normally to themselves until they contact the highest sensed points of the surfaces (Fig. 8.33). The other is to move the three coordinate system planes normally to themselves, and by half of the form deviation plus half of the orientational deviation.

In the first case a measuring strategy is needed that confines sensing to the critical areas containing the highest points. A guideline for this measuring strategy does not yet exist.

In the second case the amount of transformation calculated may be too large when the surface is convex.

However, in general, the resulting error may be negligible when a sufficient number of points are sensed. It is important that the coordinate measuring machine can achieve this transformation in order to avoid the described systematic error between datum systems according to VD&T and according to ISO 5459. Otherwise deviations of location according to ISO 1101 obtained with coordinate measuring machines will be wrong.

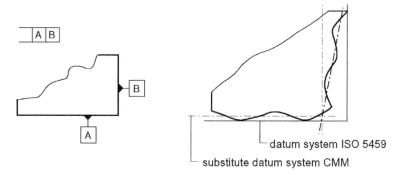

Fig. 8.33 Datum system derived from substitute elements according to VD&T and transformed to the highest sensed points

8.4   Conversion between systems and combination of systems

Both systems (the conventional system and the vectorial system) have advantages and disadvantages. For example, the vectorial system has advantages for manufacturing process control, while the conventional system has advantages for specifying the functional requirements in the cases of clearance fits.

It is likely that algorithms will be developed that for particular features will convert conventional tolerances (e.g. according to ISO 1101) into vectorial tolerances, and perhaps also vice versa.

So, in one layer of the workpiece data the functional requirements (for assembly) may be indicated by surface profile tolerances related to a workpiece datum system. In another layer of the workpiece data the vectorial tolerances (for manufacturing control) related to the same workpiece datum system may be indicated. For both indications the same nominal data (location and orientation vectors) apply. See also 19.2.

The vectorial system is normally applied in coordinate measuring machines. It is explained in the Draft German Standard DIN 32 880-1 Entwurf 1986. An equivalent ISO Standard is planned but has not up to this time been published. Depending on their software some coordinate measuring machines can also apply the conventional system.

A quick check in order to find out what the coordinate measuring machine is doing is to measure a workpiece with large form deviations, both with the coordinate measuring machine and with conventional means, e.g. with dial gauges and micrometers.

9

Maximum Material Requirement

9.1 Definitions

Actual local size (two-point size): Any individual distance at any cross-section of a feature of size (see p. xviii), i.e. any size measured between any two opposite points (two-point measurement) (ISO 286, ISO 2692) (Fig. 9.1).

Each feature of size (see p. xviii) of an individual workpiece theoretically has an infinite number of actual local sizes.

In the past, the definition of actual local size was not unambiguously standardized. The problems were the definitions of "opposite" and of the directions of the cross-sections (see 8.3.1). The new standard ISO 14 660-2 gives definitions of two-point size and various other sizes. The two-point size is the default (i.e. if not otherwise specified).

Maximum material condition (MMC): The state of the considered feature of size in which the feature is everywhere at that limit of size where the material of the feature is at its maximum, e.g. minimum hole diameter and maximum shaft diameter (ISO 2692). The actual axis of the feature need not be straight.

Mating size for an external feature: The dimension of the smallest perfect feature (e.g. cylinder or two parallel opposite planes) that can be circumscribed about the feature so that it just contacts the surface at the highest points (Fig. 9.1).

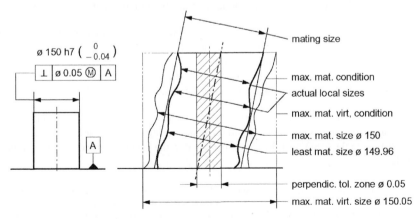

Fig. 9.1 Sizes and conditions of a feature with a related geometrical tolerance

Mating size for an internal feature: The dimension of the largest perfect feature (e.g. cylinder or two parallel opposite planes) that can be inscribed within the feature so that it just contacts the surface at the highest points (ISO 2692).

Maximum material size (MMS): The dimension defining the maximum material condition of a feature (Fig. 9.1) (ISO 2692), i.e. the limit of size where the material is at the maximum, e.g. maximum limit of size of a shaft or minimum limit of size of a hole.

Maximum material virtual size (MMVS): Size generated by the collective effect (concerning mating) of the maximum material size (MMS) and the geometrical tolerance followed by the symbol Ⓜ, i.e.

for shafts MMVS = MMS + geometrical tolerance
for holes MMVS = MMS − geometrical tolerance

The MMVS represents the design dimension of the functional gauge (ISO 2692).

Maximum material virtual condition (MMVC): A feature limiting boundary of perfect (geometrical ideal) form and of maximum material virtual size (MMVS) (Fig. 9.1). When more than one feature or one or more datum feature is applied to the geometrical tolerance, the MMVS is in the theoretical exact locations and orientations relative to each other (ISO 2692). See also 23.4.

9.2 Description of the maximum material requirement

Maximum material requirement (MMR): Requirement, indicated on drawings by the symbol Ⓜ placed after the geometrical tolerance of the toleranced feature or after the datum letter in the tolerance frame, which specifies the following:

- when applied to the toleranced feature, the maximum material virtual condition (MMVC) of the toleranced feature shall not be violated (see 9.1);
- when applied to the datum, the related maximum material virtual condition (MMVC) of the datum feature shall not be violated.

The size of the related maximum material virtual condition of the datum feature is:

- maximum material size when the datum has no geometrical tolerance followed by the symbol Ⓜ (Figs 9.2 and 9.3)* (this has the same effect as the indication Ⓔ after the size tolerance of the datum feature);
- maximum material size + (for shafts) or − (for holes) the geometrical tolerance followed by the symbol Ⓜ and applied to the datum (Fig. 9.4)* (ISO 2692).

In the case of Fig. 9.5 a symmetry tolerance applies to the datum B. However, since the tolerance frame is not connected to the datum triangle of B this tolerance does not contribute to the MMVS of the datum B related to the positional tolerance of the four holes.

---

* It is good practice to indicate the datum triangle at the tolerance frame of the tolerance applied to the datum when this tolerance is followed by the symbol Ⓜ. This indicates that this tolerance goes into the calculation of the MMVS of the datum, see Fig. 9.4. When there is no geometrical tolerance applied to the datum or a tolerance without Ⓜ the datum triangle should be indicated in extension of the dimension line, see Figs 9.2 and 9.3.

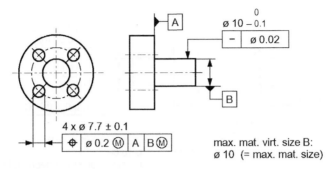

Fig. 9.2 Maximum material virtual size of datum B; form tolerance (straightness) to be disregarded

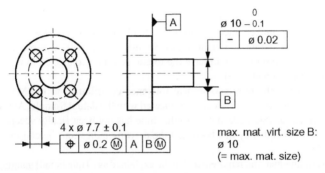

Fig. 9.3 Maximum material virtual size of datum B; form tolerance (straightness) to be disregarded

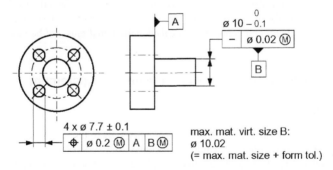

Fig. 9.4 Maximum material virtual size of datum B; form tolerance (straightness) to be regarded

For the MMVS of a datum only those geometrical tolerances come into consideration that are unrelated or related to the datums of the considered tolerance frame (Figs 9.4 and 21.9). Geometrical tolerances of the datum feature followed by Ⓜ that are related to other features do not come into consideration for the MMVS (Figs 9.5 and

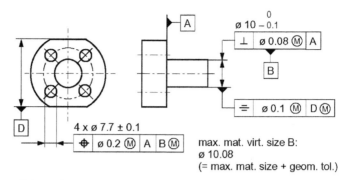

Fig. 9.5 Maximum material virtual size of datum B; perpendicularity tolerance to be regarded but symmetry tolerance to be disregarded

21.11). In order to make it more obvious which tolerances followed by Ⓜ contribute to the MMVS of the datum, the planned new version of ISO 2692 will probably specify that those that contribute should have the tolerance frame connected to the datum triangle (Figs 9.4, 9.5 and 9.15).

In other words, the only geometrical tolerance contributing to the MMVS of the datum is the tolerance frame which is connected with the datum triangle. At present this rule is planned for ISO 2692 but up to this time has not been finally accepted. In any case it is recommended by this book that the drawing indications be chosen according to this rule.

The maximum material requirement can be explained as a (functional) gauging requirement. The maximum material virtual condition at the toleranced feature and at the datum(s) describes the theoretical form, the theoretical sizes and the theoretical locations and orientations of the gauge surfaces. The workpiece must fit into this gauge. The gauge also represents the most unfavourable counterpart. When the workpiece fits into the gauge, it also fits into all counterparts. The maximum material virtual condition is the (imaginary and geometrical ideal) boundary that must not be violated.

9.3 Application of the maximum material requirement

9.3.1 General

The maximum material requirement can be applied only to those features with an axis or a median plane (cylindrical features or features composed of two opposite parallel planes) (Table 9.1). It cannot be applied to a plane surface or a line on a surface.

The maximum material requirement can be applied when there is a functional relationship between size and form or size and orientation or size and location, i.e. when the geometrical deviation may be larger if the size deviation is smaller. This applies normally to clearance fits. For transition fits and interference fits and for kinematic linkages (e.g. distances of axes of gears) the maximum material requirement is normally not appropriate. This is because enlarging the geometrical tolerance (when the size tolerance is not fully used) is detrimental to the function of the part.

Table 9.1 Possible applications of Ⓜ and Ⓛ

Tolerance		Symbol	Toleranced feature					Datum feature				
			Section line	Edge	Axis	Medion face	Surface	Section line	Edge	Axis	Medion face	Surface
Unrelated tolerance	Line profile	⌒										
	Straightness	–			X							
	Roundness	○										
	Surface profile	⌓										
	Flatness	▱				X						
	Cylindricity	⌭										
Related tolerance	Line profile with datum	⌒										
	Surface profile with datum	⌓										
	Inclination	∠			X	X				X	X	
	Parallelism	//			X	X				X	X	
	Perpendicularity	⊥			X	X				X	X	
	Position	⊕			X	X				X	X	
	Coaxiality	◎			X					X		
	Symmetry	⩵			X	X				X	X	
	Circular run-out	↗										
	Total run-out	↗↗										
Ⓜ Max. Mat. Requirement								Ⓛ Least Mat. Requirement				

9.3.2 Maximum material requirement for the toleranced feature

The maximum material requirement for the toleranced feature allows an increase in the geometrical tolerance when the feature deviates from its maximum material condition (in the direction of the least material condition), provided that the maximum material virtual condition (gauge boundary) is not violated (Figs 9.6 to 9.12).

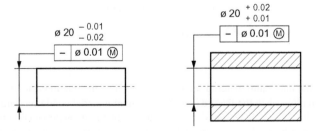

Fig. 9.6 Maximum material requirement applied to the straightness tolerance of the axis

Fig. 9.7 Gauge boundary for GD&T according to Fig. 9.6

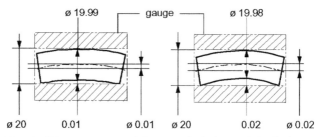

Fig. 9.8 Permissible extreme straightness deviations of the bolt according to Fig. 9.6

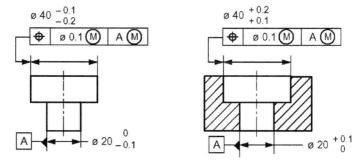

Fig. 9.9 Maximum material requirement for part and counterpart

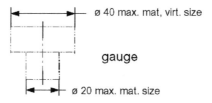

Fig. 9.10 Gauge for GD&T according to Fig. 9.9

That is, the maximum material requirement specifies that the indicated geometrical tolerance applies when the feature is in its maximum material condition (largest shaft, smallest hole). When the feature deviates from the maximum material condition (thinner shaft, larger hole), the geometrical deviation may be larger without endangering the mating capability.

The maximum material virtual condition represents the theoretically functional gauge at the toleranced feature. The maximum material virtual size represents the theoretical size of the functional gauge.

Figure 9.6 shows a bolt that is to fit into a hole. According to the definition of the maximum material requirement, in both cases (bolt and hole) the maximum material virtual size (gauge size) is ø 20 (Fig. 9.7). When the bolt is everywhere at its maximum material size ø 19.99, the straightness deviation of its axis may be 0.01, as indicated in the tolerance frame (Fig. 9.8). When the bolt is thinner, the straightness deviation may be larger. In the extreme case when the bolt is everywhere at its least material size, the straightness deviation may be 0.01 + 0.01 = 0.02 (Fig. 9.8). In any case the bolt fits into the gauge and therefore into the most unfavourable counterpart (hole). For the counterpart (hole) the corresponding applies, i.e. the gauge must fit into the hole. Therefore both parts fit.

Figure 9.9 shows two parts that are to fit together. From GD&T and according to the definition of the maximum material requirement, the maximum material virtual condition (gauge boundary) given in Fig. 9.10 is established.

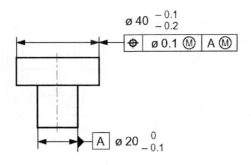

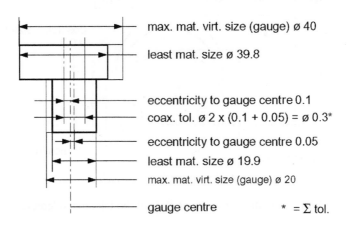

Fig. 9.11 Maximum material requirement and maximum possible coaxiality deviation

When the shafts have everywhere maximum material size ø 39.9 and ø 20 (and geometrical ideal form) the coaxiality deviation may be 0.05, which corresponds to the coaxiality tolerance of ø 0.1, as indicated in the tolerance frame (in Fig. 9.9) (see Table 18.1). When the bolt head is thinner, the coaxiality deviation may be larger. In the extreme case when the bolt head has everywhere the least material size ø 39.8 (and geometrical ideal form), the coaxiality deviation may be 0.1, which corresponds to a coaxiality tolerance of ø 0.2 (i.e. 0.1 + 0.1 = 0.2) (Fig. 9.11). In any case the bolt fits into the gauge and therefore also into the most unfavourable counterpart. For the counterpart (the two holes) the corresponding applies, i.e. the gauge must fit into the holes.

Figure 9.12 shows a pattern of holes to fit with bolts. The geometrical ideal positions of the hole axes are determined by the theoretical exact dimensions (in rectangular frames). The positional tolerance ø 0.2 of the holes is derived from the maximum material sizes, i.e. the maximum bolt ø 3 and the minimum hole ø 3.2 as described in 6.7.

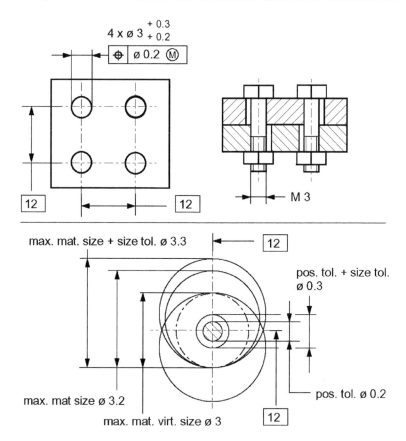

Fig. 9.12 Maximum material requirement Ⓜ at the toleranced feature

When the holes have maximum material size ø 3.2 and their actual axes are at maximum permissible separation (i.e. at the border of the tolerance zone ø 0.2 on opposite sides), the holes still provide free space for the thickest bolt. When the holes become larger than ø 3.2 (e.g. ø 3.3), the positional tolerance ø 0.2 can be increased by the amount the hole is enlarged (0.1) to become ø 0.3. This allows the actual axes of the holes to be 0.3 apart from each other (or 0.15 apart from the theoretically exact positions). In any case a gauge of maximum material virtual size and theoretical exact position fits with the holes, i.e. the hole surfaces must not violate the maximum material virtual condition (gauge boundary). This is the real criterion of the maximum material requirement.

9.3.3 Maximum material requirement for the datum

The maximum material requirement for the datum permits floating of the datum axis or datum median plane relative to the toleranced features pattern when the datum feature deviates from its maximum material condition (in the direction of the least material condition). A prerequisite is that the datum feature does not violate its maximum material virtual condition, which is geometrically ideally positioned in relation to the geometrical ideal position of the toleranced features. Within this boundary, the datum feature may take, if possible, the position where the requirements at the toleranced features are fulfilled.

The amount of float (width or diameter of the floating zone) is equal to the difference between the maximum material virtual size and the mating size of the datum feature (Fig. 9.13). This applies for primary datums as in Fig. 9.13. For secondary or tertiary datums, as in Figs 21.8 and 21.9, the amount (size of the floating zone) is smaller because of the perpendicularity deviations that must be taken into account.

The deviation of the datum feature from its maximum material virtual size does not increase the tolerance of the toleranced features relative to each other. It only permits a displacement of the pattern of tolerance zones (maximum material virtual conditions of the toleranced features) relative to the actual axis or actual median face of the datum feature (Fig. 9.13). However, when only two features are related, as in Fig. 9.11, the floating (displacement) has the effect of enlarging the tolerance of the toleranced feature.

The functional gauge embodies the maximum material virtual condition of the datum. The maximum material virtual size represents the theoretical size of the gauge. (ISO

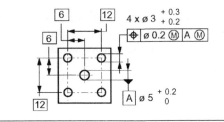

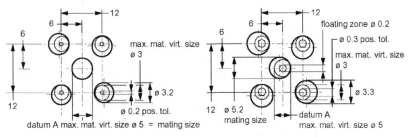

Fig. 9.13 Maximum material requirement Ⓜ for the datum

2692 - 1988 is still limited to primary datums, which have no form tolerance applied to the axis or median face. In this case the maximum material virtual condition is identical with the maximum material condition, see 9.2.)

In other words, the maximum material requirement describes a gauge contour (at the toleranced feature(s) and at the datum feature(s)) located and orientated theoretically and exactly in relation to each other, in which the actual features must be contained (the workpiece must fit into the gauge).

It should be noted that the indication of the maximum material requirement Ⓜ at the datum has a meaning different from ISO 5459. With Ⓜ at the datum the datum cylinder may float within the maximum material virtual condition (gauge). Without Ⓜ the datum cylinder is fixed.

Figure 9.11 shows for the example of Fig. 9.9 the maximum possible deviation 0.05 (eccentricity) of the datum axis relative to the gauge axis (which represents the geometrical ideal position). The maximum possible deviation of the toleranced features (head) axis relative to the gauge axis is 0.1. The maximum possible distance between both bolt axes is 0.15, corresponding to a coaxiality tolerance of ø 0.3 (see Table 18.1).

Figure 9.13 shows a pattern of holes similar to Fig. 9.12 but with a centre hole as datum. The surface of the datum hole must not violate the maximum material virtual condition (gauge boundary) of ø 5.

When the datum hole is everywhere at its maximum material size (and of geometrical ideal form), the position of the positional tolerance zones and of the maximum material virtual conditions of the four holes relative to the datum hole are determined by the theoretical exact dimension 6.

When the datum hole is larger, it can float within the boundary of the maximum material virtual condition that is represented by the gauge. The diameter of the floating zone within which the actual datum axis can float is equal to the difference between the maximum material virtual size and the mating size of the datum feature. When the mating size of the datum feature is equal to the least material size ø 5.2, the diameter of the floating zone is ø 0.2.

The theoretical exact dimension 6 determines the position of the maximum material virtual conditions (theoretical gauge dimensions). The distances between the actual datum axis and the pattern of the maximum material virtual conditions of the four holes (gauge axes at the four toleranced holes) may vary, corresponding to the floating zone. This does not alter the magnitudes of the maximum material virtual conditions (gauge dimensions) of the four holes or the distances between them. See also 23.4.

All of these requirements can be summarized by the following:

The maximum material virtual conditions of the toleranced features and of the datum feature in their theoretically exact location relative to each other determine the boundary of the functional gauge within which the part must fit.

In Fig. 9.14 the additional requirement of perpendicularity relative to the datum surface B is indicated. Accordingly, the maximum material virtual conditions of the toleranced features (four holes) and of the datum A feature (centre hole) are perpendicular to the datum B (plane surface). Here the diameter of the floating zone is not equal to the difference between the maximum material virtual size and the mating size, as in Fig. 9.13, but equal to the difference between the maximum material virtual size and the size of the contacting element perpendicular to the datum B (size of the largest inscribed cylinder perpendicular to the datum B). See also 23.4 and Fig. 20.36.

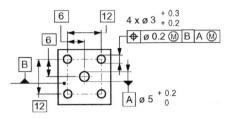

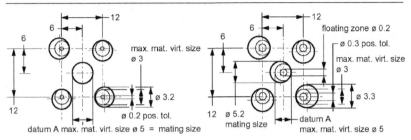

Fig. 9.14 Maximum material requirement Ⓜ for the datum, and additional requirement of perpendicularity to the plane surface datum B

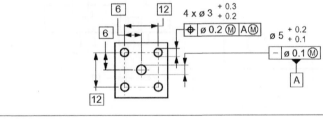

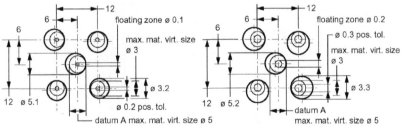

Fig. 9.15 Maximum material requirements Ⓜ for the datum and straightness tolerance to be taken into account for the datum

This means, in terms of gauging, that the gauge face must be adjusted with the workpiece datum B surface, and then the gauge mandrels must go through the holes (see 20.8.9).

In Fig. 9.15 the maximum material requirement applies to the datum feature controlling the position of the four holes. In addition, for the datum feature itself there applies a straightness tolerance for the hole axis with the maximum material requirement. The maximum material virtual condition of the feature alone (as an isolated feature) is derived from the maximum material size (ø 5.1) and the straightness tolerance (ø 0.1). Therefore the maximum material virtual size is ø 5. The same maximum material virtual condition applies to the feature as the datum A for the positional tolerance of the four holes. This is indicated by the datum triangle at the tolerance frame of the straightness tolerance of the datum feature. See also 23.4.

9.3.4 Maximum material requirement 0 Ⓜ

When the functional tolerance is not distributed on size and position, but is provided for both (size and position) for random distribution, this is to be indicated on the drawing by 0 Ⓜ (Fig. 9.16). Here also, the maximum material virtual condition must not be violated (in Fig. 9.17 the cylinder of maximum material virtual size = maximum material size = ø 3). See also 9.3.5 and 9.3.6.

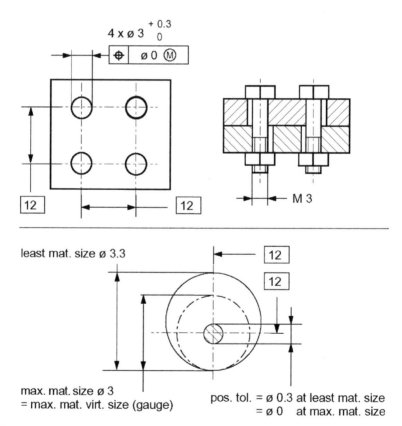

Fig. 9.16 Maximum material requirement 0 Ⓜ

Sometimes on drawings the indication according to the right-hand part of Fig. 9.17 appears. This has the same meaning as the indication using the symbol Ⓔ according to ISO 8015 and shown on the left of Fig. 9.17. In both cases the envelope requirement applies, i.e. the boundary of geometrical ideal form and maximum material size must not be violated. The symbol Ⓔ has been standardized because the drawing indication

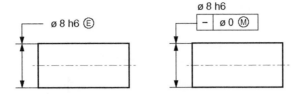

Fig. 9.17 Envelope requirement

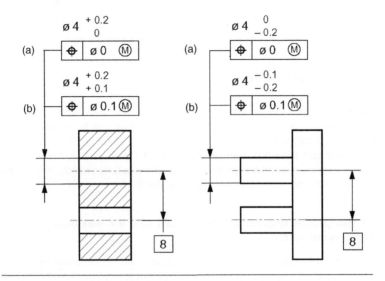

max. mat. virt. condition (gauge)

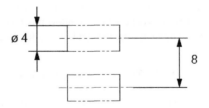

Fig. 9.18 Comparison of 0.1 Ⓜ and 0 Ⓜ with the same maximum material virtual condition (gauging boundary)

is simpler than with 0Ⓜ and in order to avoid difficulties in interpretation according to the chosen geometrical tolerance symbol (see 10).

9.3.5 Comparison of 0.1 Ⓜ and 0 Ⓜ

Figure 9.18 shows a plug and socket that shall fit. The same maximum material virtual conditions (gauging boundaries) apply to both, in cases (a) as well as in cases (b). The

max. mat. size hole ø 4.1

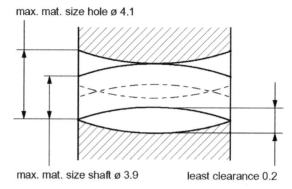

max. mat. size shaft ø 3.9   least clearance 0.2

Fig. 9.19 Least clearance (0.2) according to the tolerancing of Fig. 9.18 (b) with bent bolt and bent hole

difference between cases (a) and (b) is the size tolerance and therefore the possible clearance between bolt and hole. In case (a) the clearance may be 0 (hole ø 4 and bolt ø 4). In case (b) the clearance is at least 0.2 but may be utilized by the straightness deviations of the axes (bent hole and bent bolt) (Fig. 9.19). In case (b) the manufacturer obtains a recommendation for the distribution of the in-total provided tolerance (0.2) on the size (0.1) and on the distance (0.1) (see also 9.3.6).

9.3.6 Reciprocity requirement associated with the maximum material requirement

Figure 9.20 shows three fits (a), (b) and (c). In all cases the same maximum material virtual conditions (gauging boundaries) apply for part and counterpart. The differences between cases (a) and (b) are as described in 9.3.4 and 9.3.5. In case (b) the coaxiality tolerance will be enlarged by the size tolerance, not utilized but not vice versa. The size tolerance cannot be enlarged by the non-utilized coaxiality tolerance, although the function (clearance fit) would allow this. To allow this on the drawing, the reciprocity requirement with the symbol Ⓡ after the symbol Ⓜ after the geometrical tolerance shall be applied.

Reciprocity requirement associated with the maximum material requirement RR: Requirement, indicated on drawings by the symbol Ⓡ placed after the symbol Ⓜ after the geometrical tolerance in the tolerance frame, specifies that the maximum material virtual condition (MMVC) of the toleranced feature shall not be violated. Deviations of size, form, orientation and location may take full advantage of the total tolerance (sum of tolerances).

The reciprocity requirement associated with the maximum material requirement has the same effect as 0 Ⓜ, i.e. the total tolerance may be utilized for deviations of size, form, orientation or location in an arbitrary way. However, in contrast to 0 Ⓜ, the drawing indication with the reciprocity requirement Ⓡ gives a recommendation to the manufacturer for the distribution of the total tolerance on size and geometrical characteristics. Thus the reciprocity requirement provides communication between production planning and workshop. The design requirement (function-related tolerance) is "0 Ⓜ".

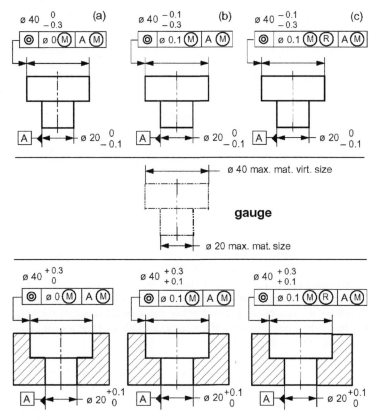

Fig. 9.20 Maximum material requirement (Ⓜ after the geometrical tolerance and after the datum letter) and reciprocity requirement (Ⓡ after Ⓜ after the geometrical tolerance)

See also 20.11 and Figs 20.48, 20.50, 20.62, 20.64 and 20.66.

For the reciprocity requirement associated with the least material requirement see 11.4.

9.4 Education

The fact that the application of the maximum material requirement appears rather complicated sometimes leads to the false opinion that application is not practicable and therefore to be ignored.

However, in many cases the precise functional requirements can only be indicated with the aid of the maximum material requirement. Only then do the largest possible tolerances appear. Therefore, often this is unavoidable when economic production is to be achieved.

A prerequisite for application of the maximum material requirement is appropriate education and appropriate planning of the manufacturing and inspection. See also 15.2 and 18.7.12.

For education, in most cases it should be sufficient to explain the maximum material requirement by the following gauging rule:

Where Ⓜ occurs, gauging is required or to be simulated. At the toleranced feature the gauge size is to be calculated in the following way from the maximum material size and the geometrical tolerance, which is followed by the symbol Ⓜ:

for shafts: maximum material size + geometrical tolerance
for holes: maximum material size − geometrical tolerance

At the datum feature the gauge size is to be calculated in the same way when at the datum feature a geometrical tolerance followed by the symbol Ⓜ is indicated (and the datum triangle is connected directly to this tolerance frame) (Figs 9.15 and 9.4).

When there is no geometrical tolerance followed by the symbol Ⓜ indicated at the datum, the gauge size is equal to the maximum material size (Figs 9.2, 9.3 and 9.14).

Designers should know that at the datum (of the original geometrical tolerance) only geometrical tolerances (tolerance frames) should be connected with the datum triangle that have no relationship to other features (straightness of an axis, Fig. 9.4) or that are related to datums occurring in the datum system of the original tolerance (Fig. 9.5).

According to ANSI Y14.5M - 1982 and to ASME Y14.5M - 1995 there are different rules. Connection of the datum triangle to the considered tolerance frame is not mandatory. The standard requires an analysis of tolerance controls applied to a datum feature in determining the size of the gauge. Further, the standard specifies rules for cases when at the datum feature geometrical tolerances are indicated that are not followed by the symbol Ⓜ. If necessary see 21.1.2.

10

Envelope Requirement

10.1 Definition

The envelope requirement according to ISO 8015 specifies that the surface of a single feature of size (cylindrical surface or a feature established by two parallel opposite plane surfaces) should not violate the imaginary envelope of perfect (geometrical ideal) form at maximum material size.

The envelope requirement may be specified either

- by indication of the symbol Ⓔ placed after the linear (size) tolerance, when applicable to a selected individual feature;
- by indication in the drawing title box "ISO 2768 ... -E", when applicable to all features of size in addition to the general geometrical tolerances according to ISO 2768;
- by a national standard, e.g. ASME Y14.5M or DIN 7167, when applicable to all features of size.

The envelope requirement cannot be applied to features for which a straightness or flatness tolerance is specified that is larger than the size tolerance. According to ASME Y14.5M, the envelope requirement also does not apply to features for which a straightness tolerance of the axis (even if smaller than the size tolerance) is specified.

The envelope requirement may also be indicated by a form tolerance with 0 Ⓜ applied, see 9.3.4. However, it is rather difficult to find the proper drawing indication (Fig. 10.1). Therefore the symbol Ⓔ has been standardized in ISO 8015.

10.2 Application of the envelope requirement

The envelope requirement is applicable to single features of size (cylindrical surfaces or features established by two parallel opposite plane surfaces).

The envelope requirement may be applied to features of size that are to be mated with a clearance fit.

Figure 10.2 shows the application of the envelope requirement for a cylindrical feature. The size tolerance requires that the actual local sizes are within the limits of size, see 8.3.1. In addition the envelope requirement is specified.

When the envelope requirement is not specified, the actual local sizes within the cross-section could be within the limits of size; however, the cross-sections could have a lobed form and go beyond the circles of maximum material size by the amount of the roundness tolerance (e.g. general roundness tolerance) (Fig. 10.3). In addition, the feature could

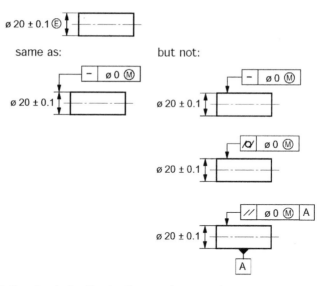

Fig. 10.1 Drawing indication for the envelope requirement

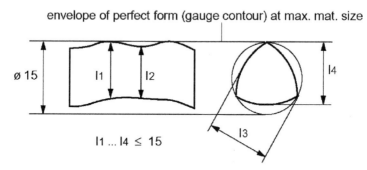

Fig. 10.2 Envelope requirement

go beyond the envelope of perfect form at maximum material size by the amount of the straightness tolerance (e.g. general straightness tolerance) even if the feature is everywhere at maximum material size (Fig. 10.4). In other words, the feature may violate the envelope requirement even when the size tolerance and the form tolerance are respected. Therefore inspection of the size deviation and of the form deviation alone is not sufficient to verify the envelope requirement.

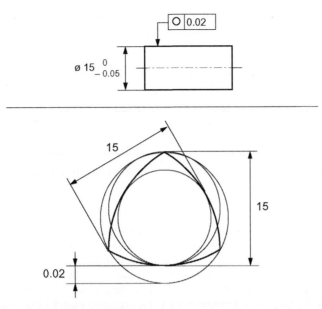

Fig. 10.3 Violation of the envelope requirement even though size and roundness tolerances are respected

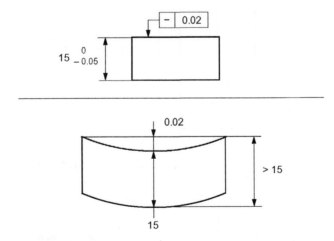

Fig. 10.4 Violation of the envelope requirement even though size and straightness tolerances are respected

In order to verify the envelope requirement, gauging or a gauging simulation, e.g. on a coordinate measuring machine, must be executed (see 18.7.11), and the manufacturing size tolerance must be diminished at the maximum material side by the amount of the expected form deviations.

Figure 10.5 shows an example where the envelope requirement applies over a length of 10 but throughout the whole cylinder. For example, when a ring of 10 width moves along the cylinder.

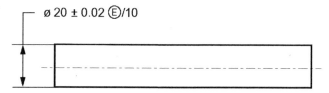

ø 20 ± 0.02 Ⓔ/10

Fig. 10.5 Envelope requirement over a restricted length lying anywhere

10.3 Cross-sections within size tolerance fields

The envelope requirement deals with three-dimensional features. Sometimes, however, specifications deal with the circumference lines in cross-sections (e.g. in the German Standards DIN 1748 and DIN 17615 on aluminium sections). The specification requires the circumference lines in cross-sections normal to the axis to be contained between limiting lines that are geometrical ideal, concentric, in the geometrical ideal orientation relative to each other, and of distances equal to the maximum material sizes and the least material sizes (Fig. 10.6).

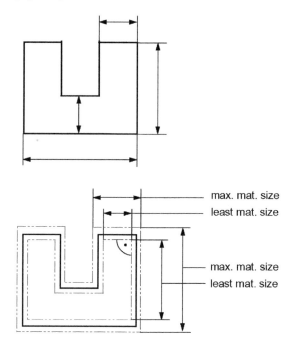

max. mat. size
least mat. size

max. mat. size
least mat. size

Fig. 10.6 Cross-sections between concentric geometrical ideal limiting lines of geometrical ideal orientation and of distances equal to the maximum material sizes and least material sizes

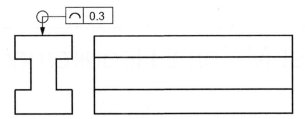

Fig. 10.7 Line profile tolerance around each cross-section

This specification differs from the envelope requirement as follows:

- the entire feature is not considered, but cross-sections only, and each cross-section is taken to be independent of the others;
- not only is the geometric ideal form at maximum material size specified, but also that at least material size;
- the specification is not limited to single lines but to all lines establishing the entire circumference and taking into account the geometrical ideal orientation and location of the line elements relative to each other.

Symbols for this specification are not standardized.

The form deviations along the axis (straightness tolerance, flatness tolerance) are to be specified separately. DIN 1748 and DIN 17 615 specify the twist tolerance (see 3.7) instead of a flatness tolerance.

When the tolerance zone is equal around the section, a line profile tolerance around may be specified (Fig. 10.7).

11

Least Material Requirement

11.1   Definitions

Least material condition (LMC): The state of the considered feature in which the feature is everywhere at that limit of size where the material of the feature is at its minimum, e.g. maximum limit of size of a hole (maximum hole diameter) and minimum limit of size of a shaft (minimum shaft diameter) (ISO 2692).

The actual axis of the feature need not be straight.

Least material size (LMS): The dimension defining the least material condition of a feature (ISO 2692), i.e. the limit of size where the material is at the minimum (e.g. maximum limit of size of a hole and minimum limit of size of a shaft).

Least material virtual size (LMVS): Size generated by the collective effect (with regard to what can be cut out of the material) of the least material size and the geometrical tolerance followed by the symbol Ⓛ, i.e.

for shafts: LMVS = LMS − geometrical tolerance
for holes: LMVS = LMS + geometrical tolerance

Least material virtual condition (LMVC): The features limiting boundary of perfect (geometrical ideal) form and of least material virtual size. When more than one feature, or one or more datum features are applied to the geometrical tolerance, the LMVS are in theoretical exact locations and orientations relative to each other (ISO 2692).

11.2   Description of least material requirement

Least material requirement (LMR): This requirement, indicated on drawings by the symbol Ⓛ placed after the geometrical tolerance of the toleranced feature or after the datum letter in the tolerance frame, specifies:

- when applied to the toleranced feature, the least material virtual condition (LMVC) of the toleranced feature shall be fully contained within the material of the actual toleranced feature;
- when applied to the datum, the least material virtual condition (LMVC) of the datum feature shall be fully contained within the material of the actual datum feature (ISO 2692).

The least material virtual condition at the datum feature is usually at least material size, because usually at the datum feature no geometrical tolerance followed by the symbol Ⓛ is indicated.

The least material requirement (LMR) Ⓛ has the effect that the least material virtual condition is entirely contained in the material of the feature and, for example, can be cut out. The mutual dependence of size and form or location and orientation is thereby taken into consideration.

A typical application is a casting when the final machined part shall be achievable.

11.3   Application of least material requirement

11.3.1 General

The least material requirement can be applied only to those features that have an axis or median plane (cylindrical features or features composed of two parallel opposite planes) (Table 9.1).

The least material requirement may be applied where a minimum material thickness must be respected and this minimum thickness depends on the mutual effect of deviations of size and form, location or orientation.

11.3.2 Least material requirement for the toleranced feature

The least material requirement for the toleranced feature allows an increase in the geometrical tolerance when the feature deviates from its least material condition (in the direction of the maximum material condition), provided that the least material virtual condition is not violated (is entirely within the material) (Fig. 11.1).

That is, the least material requirement specifies that the indicated geometrical tolerance applies when the feature is in its least material condition (smallest shaft, largest hole) and of geometric ideal form. When the feature deviates from the least material condition (larger shaft, smaller hole) the geometrical deviation may be larger without violating the least material virtual condition.

Figure 11.1 shows an example where a minimum ridge thickness shall be contained in the material, e.g. of a casting that is to be machined, and Fig. 11.2 shows a permissible

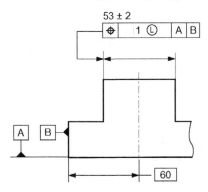

Fig. 11.1 Least material requirement

workpiece. The least material virtual condition has the size of 50 (least material size minus positional tolerance), is perpendicular to the datum (surface) A and 60 apart from the datum (surface) B. The more the ridge thickness deviates from the least material size, the more the actual median face may deviate from the theoretical exact location. In the example the smallest actual local size is 53. In this case the positional deviation of the actual median face may be 1.5, which corresponds to a positional tolerance of 3 (see Table 18.1).

Figure 11.3 shows an example where the geometrical ideal form of least material size shall be contained in the material (must be possible to cut out). Here the least material virtual condition has least material size.

Without this indication, the feature may be bent or may be of a lobed form in the cross-sections (here within the general tolerances of straightness and roundness according to ISO 2768-H). In the extreme most unfavourable case the maximum (geometrical ideal) cylinder contained in the material (possible to cut out) has a diameter of least material size minus the sum of straightness tolerance and roundness tolerance (Fig. 11.4). The indication according to Fig. 11.3 excludes this diminishing of the maximum contained cylinder.

11.3.3 Least material requirement for the datum

The least material requirement for the datum permits floating of the datum axis or datum median plane relative to the toleranced feature(s) when the datum feature deviates from its least material condition (in the direction of the maximum material condition). A prerequisite is that the surface of the datum feature does not violate the least material virtual condition (which is geometrically ideally positioned in relation to the geometrical ideal position of the toleranced features). Around this boundary, the datum

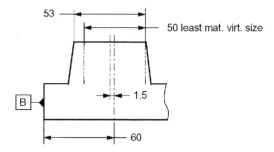

Fig. 11.2 Permissible workpiece according to the tolerancing in Fig. 11.1

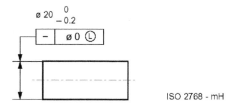

Fig. 11.3 Straightness tolerance of the axis with least material requirement

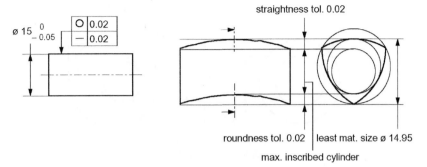

Fig. 11.4 Maximum inscribed cylinder of diameter d_{max} in a workpiece of maximum permissible straightness deviations and with lobed forms of least material size and maximum permissible roundness deviations

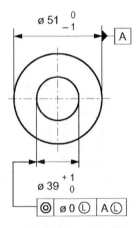

Fig. 11.5 Coaxiality tolerance with least material requirement in order to secure the minimum wall thickness

feature may take, the position where the requirements of the toleranced features are fulfilled.

It should be noted that the indication of the least material requirement Ⓛ at the datum has a meaning different from ISO 5459. With A Ⓛ the datum cylinder A of least material (virtual) size must be contained in the material (and is not allowed to lie outside of the material.

Figure 11.5 shows an example where a geometrical ideal pipe with a minimum wall thickness of 5 shall be contained in the material (must be possible to cut out). Figure 11.6 shows the coaxial least material virtual cylinders that must not be violated by the surfaces. On the left-hand side both surfaces (features) have least material size and therefore shall be coaxial. On the right-hand side are shown the extreme permissible coaxiality deviations with maximum material sizes.

See also 20.10.4.

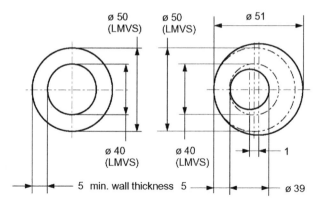

Fig. 11.6 Calculation of the minimum wall thickness 5 according to tolerancing according to Fig. 11.5; left with least material sizes; right with maximum material sizes

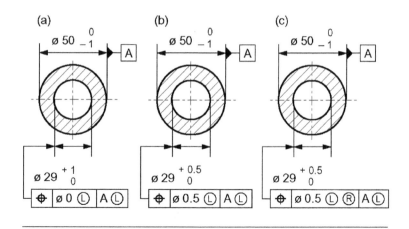

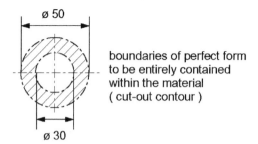

boundaries of perfect form
to be entirely contained
within the material
(cut-out contour)

Fig. 11.7 Least material requirement (Ⓛ after geometrical tolerance and after datum letter) and reciprocity requirement (Ⓡ after Ⓛ after geometrical tolerance)

11.4 Reciprocity requirement associated with least material requirement

Figure 11.7 shows three rings (a), (b) and (c). In all cases the same least material virtual conditions (boundaries to be entirely contained within the material) apply. In case (a) the whole tolerance is indicated at the size. The tolerance may be utilized by size deviations and by coaxiality deviations in an arbitrary way. In case (b) the tolerance is distributed on size and coaxiality. The coaxiality tolerance will be enlarged by the size tolerance not utilized, but not vice versa. The size tolerance cannot be enlarged by the non-utilized coaxiality tolerance. In case (c) the size tolerance can also be enlarged by the non-utilized coaxiality tolerance.

Reciprocity requirement (RR) associated with the least material requirement: This requirement, indicated on drawings by the symbol Ⓡ placed after the symbol Ⓛ after the geometrical tolerance in the tolerance frame, specifies that the least material virtual condition (LMVC) of the toleranced feature shall not be violated. Deviations of size, form, orientation and location may take full advantage of the total tolerance (sum of tolerances).

The reciprocity requirement has the same effect as 0 Ⓛ, i.e. the total tolerance may be utilized for deviations of size, form, orientation or location in an arbitrary way. However, in contrast to 0 Ⓛ, the drawing indication with the reciprocity requirement Ⓡ gives a recommendation to the manufacturer for the distribution of the total tolerance on size and geometrical characteristics. Thus the reciprocity requirement provides a communication between production planning and workshop (manufacturing-related tolerance). The design requirement (functional-related tolerance) is "0 Ⓜ ". There may be several manufacturing-related tolerances (with Ⓡ) derived from the same functional-related tolerance according to the needs of different workshops.

12

Tolerancing of Flexible Parts

Certain flexible (non-rigid) parts made, for example, from thin sheet metal, fibreglass, plastics or rubber, when removed from their manufacturing environment, deform substantially from their manufactured condition (geometrical shape) by virtue of their weight and flexibility, or by the release of internal stresses resulting from the manufacturing process.

When in the drawing, in or near the title box "ISO 10 579-NR" is indicated, all geometrical tolerances not associated with the symbol Ⓕ apply in the restrained (assembled) condition. The restrained (assembled) condition is to be defined on the drawing, examples (see Figs. 12.1, 12.2, 20.131 and 20.132).

Geometrical tolerances followed by the symbol Ⓕ apply in the free state.

For the free state the conditions should be indicated under which the geometrical tolerance under free state is ensured, e.g. the direction of gravity and, if necessary, the setting condition (support) of the part.

According to ISO 10 579, for the restrained condition, only those pressures and forces may be applied that can be expected under normal assembly conditions.

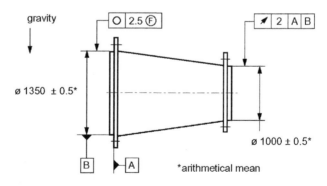

ISO 10 579-NR
Restrained condition:
Surface A is mounted with 64 bolts M6 × 1,
screwed with a torque of 9 to 15 Nm
and the B feature is restrained to the
corresponding maximum material limit

Fig. 12.1 Tolerancing of a non-rigid (flexible) part according to ISO 10 579

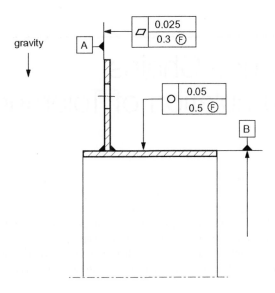

ISO 10 579-NR
Restrained condition:
Surface A is mounted with 120 bolts M 20,
screwed with a torque of 18 to 20 Nm and
the B feature is restrained to the
corresponding maximum material limit

Fig. 12.2 Tolerancing of a non-rigid (flexible) part according to ISO 10 579

When the drawing has referenced ISO 10 579-NR, the rules of this standard apply irrespective of whether the part is flexible or not.

See also 20.12.

13

Tolerance Chains
(Accumulation of Tolerances)

When workpieces are fitted together, size deviations accumulate. In order to assess the resulting clearance or interference, tolerance line-up calculations are performed. Arithmetical tolerance calculations are based on extreme cases (worst cases) when all sizes are at their favourable or unfavourable limit of size. Statistical tolerance calculations take into account the form of distribution of the actual sizes and give the clearance or interference that will not be exceeded with a certain statistical probability (see 14).

The procedure for an arithmetical tolerance line-up calculation is as follows:

1. define the dimension scheme, showing all dimensions and their tolerances that form the chain, i.e. all dimensions that contribute to the clearance or interference;
2. dimensions whose upper limits lead to an increase in the closing dimension are drawn in the positive direction, and others in the negative direction;
3. the arithmetical sum of the maximum limits of size of the positive chain links and the minimum limits of size of the negative chain links, gives the **maximum value of the closing dimension** (maximum clearance, minimum interference). See Fig. 13.1.

The arithmetical sum of the minimum limits of size of the positive chain links and the maximum limits of size of the negative chain links, gives the **minimum value of the closing dimension** (minimum clearance, maximum interference) (Fig. 13.1).

Instead of the limits of size, the permissible deviations may be taken when the arithmetical sum of the nominal sizes is zero.

The tolerance line-up calculation according to Fig. 13.1 considers size tolerances only. A prerequisite is that the workpiece surfaces respect the geometrical ideal boundaries of maximum material size. In the example of Fig. 13.2 this is the boundary consisting of a cylinder of minimum size of the holes and of two parallel planes the minimum distance apart, and being perpendicular to the cylinder.

This prerequisite does not exist according to ISO 8015. Therefore the parts must be toleranced appropriately (Fig. 13.2).

The tolerancing of the perpendicularity deviation (to be within the size tolerance) may also be chosen according to Fig. 13.3. The meaning (i.e. the go gauge) of both Figs 13.2 and 13.3 is the same. In Fig. 13.3 the indication "A Ⓜ–B Ⓜ" has the same effect as the indication "CZ" in Fig. 13.2.

A similar requirement exists for the gear as shown in Fig. 13.4.

When the perpendicularity deviations are not toleranced by "0 Ⓜ" the workpieces may have deviations as shown in Fig. 13.5.

The tolerance calculation as described in Fig. 13.1 is then not sufficient. The perpendicularity deviation may override the minimum clearance, and the assembly may jam.

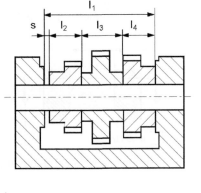

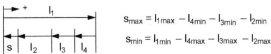

$$S_{max} = l_{1max} - l_{4min} - l_{3min} - l_{2min}$$

$$S_{min} = l_{1min} - l_{4max} - l_{3max} - l_{2max}$$

Fig. 13.1 Arithmetical tolerance line-up calculation

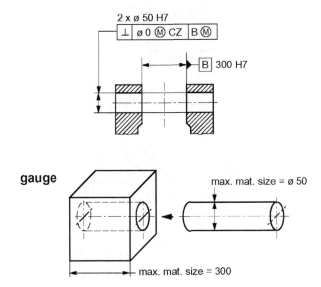

Fig. 13.2 Perpendicularity deviations within size tolerances

The tolerance calculation must then include the geometrical tolerances (Fig. 13.6).

It should be noted that the perpendicularity tolerance t_{r1} of the faces of the dimension l_1 decreases the clearance, and must therefore enter the dimension scheme in the negative direction (this in opposition to the dimension l_1, itself) (Fig. 13.8).

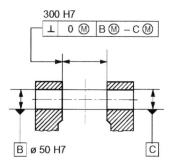

Fig. 13.3 Perpendicularity deviations within size tolerances

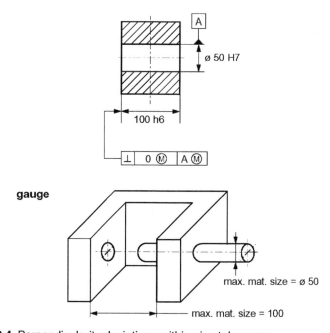

Fig. 13.4 Perpendicularity deviations within size tolerances

The perpendicularity tolerances are to be indicated in the drawing together with the maximum material requirement (e.g. according to Figs 13.7 and 13.9). Because this is an assembly with clearance, the maximum material requirement is appropriate to the function. It allows larger perpendicularity deviations when the workpiece sizes deviate from the maximum material size (approach the minimum material size).

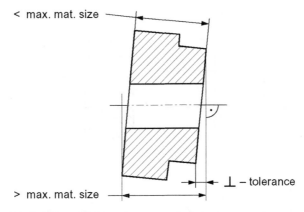

Fig. 13.5 Workpiece with maximum material sizes and perpendicularity deviations

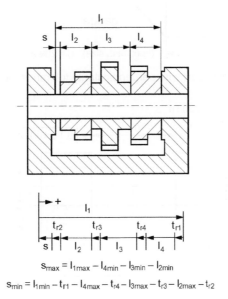

$$S_{max} = l_{1max} - l_{4min} - l_{3min} - l_{2min}$$

$$S_{min} = l_{1min} - t_{r1} - l_{4max} - t_{r4} - l_{3max} - t_{r3} - l_{2max} - t_{r2}$$

Fig. 13.6 Arithmetical tolerance line-up calculation including the perpendicularity tolerances t_r Ⓜ

Figure 13.10 shows an assembly and the tolerance line-up calculation, taking into account the perpendicularity tolerances of the side faces. In this case the assembly may jam although the tolerances are respected, because the right face has a smaller diameter than the adjacent part (Fig. 13.11). Tolerancing taking account of the size of the counterpart is appropriate to the function (Fig. 13.12).

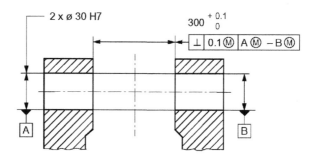

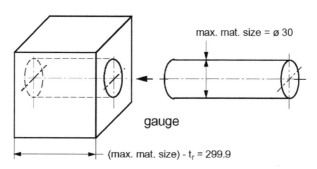

Fig. 13.7 Assembly part with perpendicularity tolerance indicated

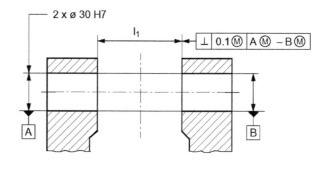

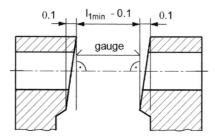

Fig. 13.8 Perpendicularity deviation of the faces reduces the gap

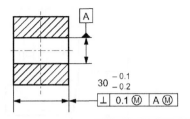

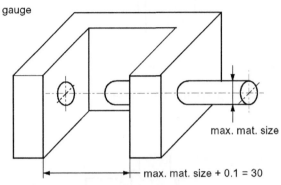

Fig. 13.9 Assembly part with perpendicularity tolerance indicated

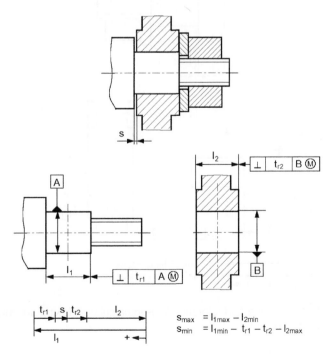

Fig. 13.10 Tolerancing and tolerance line-up calculation taking into account the perpendicularity tolerances t_r

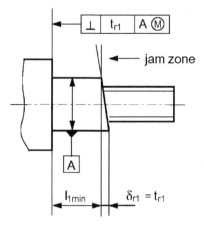

Fig. 13.11 Permissible position of the side face

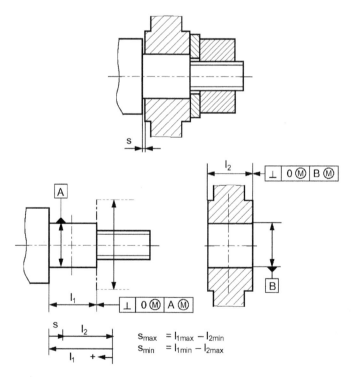

Fig. 13.12 Tolerancing and tolerance line-up calculation taking account of the size of the counterpart

14

Statistical Tolerancing

With the arithmetical tolerance line-up calculation (worst case tolerancing), the tolerances of the links of the dimension chain (e.g. single parts of an assembly) are defined in such a way that the assembly still functions when all links of the chain have actual sizes equal to their limits of size (Fig. 14.1).

However, in practice the actual sizes are subject to variations, and are therefore statistically distributed (e.g. according to a normal distribution) (Figs 14.2 and 14.3). With this

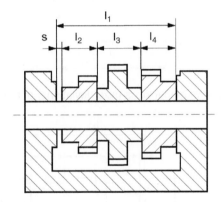

Fig. 14.1 Assembly with required clearance *s* (functional closing tolerance *T*)

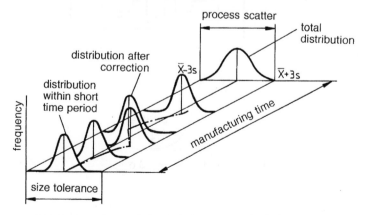

Fig. 14.2 Distribution of actual sizes during a manufacturing process

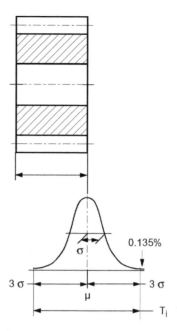

Fig. 14.3 Distribution of actual sizes (size deviations) of a part

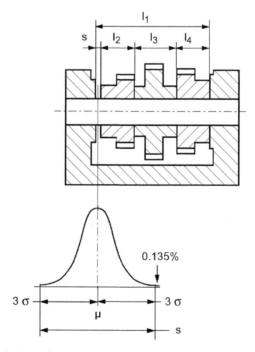

Fig. 14.4 Distribution of actual values of the closing dimension (clearance)

supposition, it is most unlikely that an assembly contains only parts with actual sizes equal to the limits of size (Fig. 14.4).

In contrast to the arithmetical tolerance line-up calculation, in the statistical tolerance line-up calculation it is assumed that this extreme case will not occur. The single tolerances are chosen larger than with arithmetical tolerancing.

The variability Δs of the clearance or interference, i.e. functional closing tolerance T of the assembly derives from n single tolerances with arithmetical tolerancing according to

$$\Delta_s = T = T_a = T_1 + T_2 + \cdots + T_n$$

with statistical tolerancing according to

$$\Delta_s = T = T_s = \sqrt{(T_{s1}^2 + T_{s2}^2 + \cdots + T_{sn}^2)}$$

With an assembly chain of four parts (Fig. 14.5), the following applies:

With arithmetical tolerancing, the single tolerance is $T/4 = 0.25T$. With statistical tolerancing, the single tolerance is $T/\sqrt{4} = 0.5T$. Non-functioning occurs when the arithmetical sum of the actual deviations exceeds the closing tolerance. The probability for this is, with statistical tolerancing (variation of the single dimension equal to one sixth

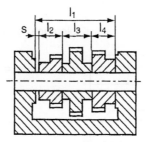

arithmetical tolerancing: single tolerance $T_{ai} = \Delta s/4 \quad = 0.25\,\Delta_s$
statistical tolerancing: single tolerance $T_{ai} = \Delta s/\sqrt{4} = 0.5\,\Delta_s$

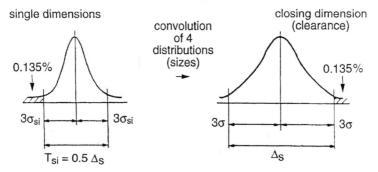

Fig. 14.5 Probability of exceeding the statistical closing tolerance of a chain of four parts. Probability of exceeding $T_s = \Delta_s$: 0.135%, i.e. 1.35 of 1000 assemblies

of the statistical tolerance and variation of the closing dimension equal to one sixth of the closing tolerance)

$$P = 0.135\%,$$

i.e. 1.35 of 1000 assemblies may jam, and at least one part must be changed for another one.

Here it is assumed that the actual sizes are normally distributed, their standard deviation is one sixth of the statistical single tolerance, the distribution is centred in the tolerance and the (dimensions of the) parts are independent of each other.

Introducing a coverage factor c covers deviations from these assumptions

$$T_s = \sqrt{\Sigma \, c^2 \, T_{si}^2}$$

Depending on the manufacturing method (form of distribution) and according to the experiences of the manufacturer, the coverage factor c ranges practically between 1.3 and 1.7.

In Ref. [2] coverage factors c are indicated, depending on the form of distribution of the single sizes (equal, trapeze, triangular, normal). Figure 14.6 shows coverage factors c often used in industrial practice (the practically worst case $c = 1.73$ occurs very seldom).

According to the central limit theorem of mathematical statistics, in any case, independent of the kind of distribution of the single sizes, the closing dimensions are approximately normally distributed if more than four members contribute to the closing dimension.

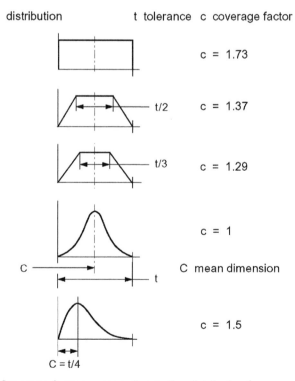

Fig. 14.6 Coverage factors c according to the distribution form

In order to stipulate and verify the assumptions of statistical tolerancing, properly defined terms are needed. An International standard on this subject does not yet exist. The German Standard DIN 7186 defines terms as follows:

Mean size C: Arithmetical mean of the limits of size.

Dimension chain: Geometrical representation of several coeffecting dimensions that are independent of each other (Fig. 14.7).

Single dimension: Dimension acting as a link in a dimension chain.

Closing dimension: Result of arithmetic addition of an independent single dimension within a dimension chain.

Single tolerance T_i: Tolerance of a single dimension.

Statistical tolerance T_{si}: Single tolerance with specifications regarding the distribution of the actual sizes within the tolerance zone.

Arithmetical closing tolerance T_a: Sum of the single tolerances within a dimension chain.

Statistical closing tolerance T_s: Closing tolerance, smaller than T_a, specified according to the actual size distribution.

Side zones and central zone: For the specification of statistical tolerances the tolerance zone is divided into zones. In general, three zones predominantly symmetrical with respect to the mean size C are sufficient (Fig. 14.8).

The lower side zone B_u is the zone adjacent to the minimum limit of size.
The upper side zone B_o is the zone adjacent to the maximum limit of size.
The central zone B_m is the zone between the upper and the lower side zone.

Side contents P_u and P_o: Percentage of actual sizes (more precise: mating sizes or related mating sizes) of a manufacturing lot within the side zone (Fig. 14.8).

If not otherwise specified, according to DIN 7186 the contents P_u and P_o must not be more than $(100\% - P_m\%)/2$ each.

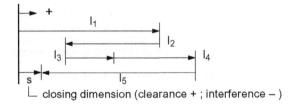

Fig. 14.7 Dimension chains

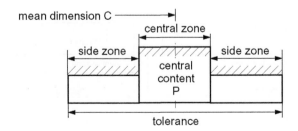

Fig. 14.8 Statistical tolerancing and tolerance zones

(From this it follows that, with an asymmetrical distribution of the actual sizes where there is no lower side zone, the content that would go into this zone actually goes into the central zone (Fig. 14.10).)

Central content *P*: Percentage of actual sizes (more precise: mating sizes or related mating sizes) of a manufacturing lot within the central zone (Fig. 14.8).

For **machining processes (chip removal)** the following specifications are predominant:

• central zone (width) and side zones (widths) each 1/3 of the tolerance T_{si};
• central content at least 50%, side contents not more than 25% each.

Figure 14.9 shows the drawing indication according to DIN 7186.

Deviations of form, orientation and run-out are not symmetrically distributed. The following specifications are predominant. Coverage factor $c = 1.5$; central zone equal to half the tolerance wide and adjacent to zero located; central content at least 75% (the left side zone is contained within the central zone). Figure 14.10 shows the two predominant possibilities of central zones.

An international standard on drawing indications for statistical tolerances has not, thus far, been published. The planned standard will probably adopt the symbol "$\langle ST \rangle$"

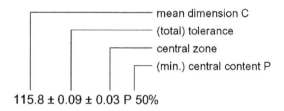

$$115.8 \pm 0.09 \pm 0.03 \text{ P } 50\%$$

(total) tolerance = max. dim. − min. dim.
(min.) central content P = min. percentage of actual dim. within central zone
(e.g. min. 50% within ± 0.03, max. 25% within each side zone)

Fig. 14.9 Statistical tolerancing, drawing indication and interpretation according to DIN 7186

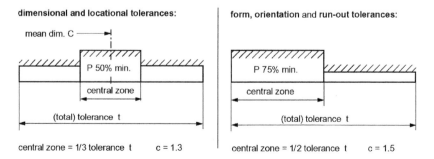

Fig. 14.10 Central zones for statistical tolerancing

of the American standard ASME Y14.5 followed by boxed specifications regarding the distribution of the specified property (e.g. size). Figure 14.11 shows some possibilities. The upper three examples specify central zones and central contents. The other examples specify process capability parameters Cp, Cpk, Cc, Cpu, Cpm which have to be defined in referenced specifications and which determine limits of the distribution of the specified property.

Figure 14.12 shows an example of an assembly. With arithmetical tolerancing, the tolerance for the dimensions $l_1 \ldots l_7$ is 0.05 (± 0.025). With statistical tolerancing, the

central tolerance zone content

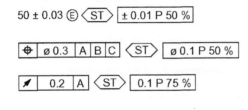

process capability indices

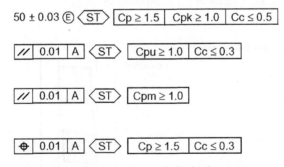

Fig. 14.11 Drawing indications for statistical tolerances

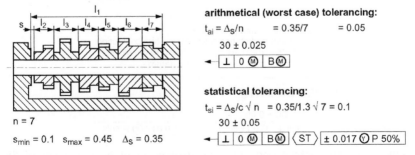

arithmetical (worst case) tolerancing:

$t_{ai} = \Delta_S/n \qquad = 0.35/7 \qquad = 0.05$

30 ± 0.025

statistical tolerancing:

$t_{si} = \Delta_S/c \sqrt{n} \; = 0.35/1.3 \sqrt{7} = 0.1$

30 ± 0.05

Fig. 14.12 Example of an assembly arithmetical toleranced and statistical toleranced under the condition of the same limits for the clearance (comparison)

tolerance is 0.1 (±0.05). Figure 14.13 shows tolerancing of the parts of the assembly. The symbol $\text{\textcircled{Y}}$ indicates related mating sizes for the dimensions $l_1 \ldots l_7$, which must be explained on the drawing because this symbol is not yet standardized.

Statistical tolerancing allows specifications of larger single tolerances than arithmetical tolerancing without detriment to the function of the workpiece. In many cases this results in a gain in manufacturing economy. Because of the current trend for greater miniaturization and more precise products, statistical tolerancing has become more important. Sometimes the smaller arithmetical tolerances are not even achievable.

However, the following prerequisites should be for manufacturing:

- the actual sizes of a single dimension are approximately normally distributed or deviations from the normal distribution are taken care of by the coverage factor c;
- the mean values of the symmetrical distributions coincide approximately with the mean size;
- the ratios of single tolerance T_{si} and standard deviation σ are approximately of the same order of magnitude, e.g. $T_{si}/\sigma = 6$;
- there is no mutual dependence between the dimensions within the dimension chain.

Statistical tolerancing is more advantageous:

- the greater the number of members in the dimension chain;
- the greater the manufacturing lot;
- the better manufacturing and inspection can satisfy the prerequisites for statistical tolerancing (form and location of the distribution, see above).

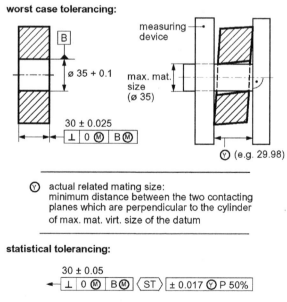

worst case tolerancing:

B

ø 35 + 0.1

30 ± 0.025

⊥ 0 Ⓜ B Ⓜ

measuring device

max. mat. size (ø 35)

Ⓨ (e.g. 29.98)

Ⓨ actual related mating size:
minimum distance between the two contacting planes which are perpendicular to the cylinder of max. mat. virt. size of the datum

statistical tolerancing:

30 ± 0.05

⊥ 0 Ⓜ B Ⓜ ⟨ST⟩ ± 0.017 Ⓨ P 50%

Fig. 14.13 Arithmetical and statistical tolerancing of a component of the assembly

Sizes produced by punching have usually very small variability within a delivery lot, i.e. the distribution is very small. For example, with the size of punched holes when the tool is new the distribution starts near the least material size of the hole. Due to the wear of the tool over the years, the distribution drifts towards the maximum material size of the hole. As for a delivery lot, normally it is not known where within the tolerance the distribution is located, the whole tolerance must be taken into account for the statistical line- up calculation (Fig. 14.14).

resulting clearance, interference $\Delta s = \Sigma T_p + \sqrt{\Sigma T_r^2}$

T_r tolerance of dimension produced by chip removal

T_p tolerance of dimension produced by punching

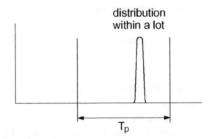

Fig. 14.14 Statistical tolerancing with chain links produced by punching

There are computer programs available for two- or three-dimensional statistical tolerance line-up calculations. These programs use methods of assembly simulation (Monte Carlo method) or convolution of distributions and facilitate considerable statistical tolerancing.

In mass production, statistical tolerancing allows more realistic tolerance calculations, leading to larger tolerances with lower production costs without loss in product quality.

15

Respecting Geometrical Tolerances during Manufacturing

15.1 Manufacturing influences

Geometrical deviations are influenced by the following, which are sometimes referred to as the "5 Ms".

Material
- rigidity of the workpiece (shape);
- material;
- stress in the material.

Machine (tool)
- precision of the machine tool, bearing play;
- static and dynamic rigidity of the machine tool;
- thermal properties of the machine tool;
- maintenance;
- environment (e.g. vibrations).

Method
- tool;
- chuck, fixing, clamping method;
- processing data (e.g. cutting speed, thickness of cut, cutting pressure, cooling).

Measuring
- uncorrected systematic measuring deviations;
- random measuring deviations.

Manufacturer
- education, skillness, precision of re-chucking;
- environment.

The following figures show some results of investigations in the field of metal removal processes. These results are only examples, and are not general.

Figure 15.1 shows the mean values $\bar{x}$ and the variations of measured roundness deviations of workpieces out of steel, ø20 mm, manufactured with certain machine tools.

Figure 15.2 shows the ability of manufacturing devices to respect tolerances of roundness, cylindricity and coaxiality by turning and grinding.

The machine tools are to be classified according to their abilities. The abilities related to the workpiece material and shape and process data (e.g. cutting speed, thickness of cut and chucking method) are to be recorded for manufacturing planning.

Figure 15.3 shows, as an example, the typical dependence of roundness deviation on cutting speed.

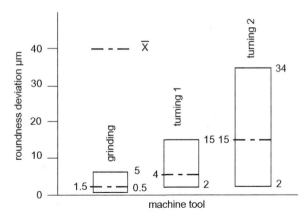

Fig. 15.1 Ability of machine tool, mean values and variations of roundness deviations of workpieces made out of steel ø20 mm

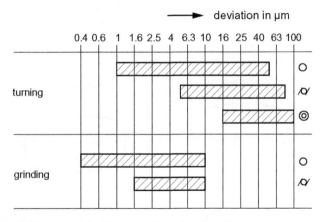

Fig. 15.2 Ability of manufacturing devices in a particular workshop

Figure 15.4 shows the result of chucking and cutting pressure influence on the workpiece form.

The largest geometrical deviations occur in general as a result of re-chucking the workpiece. Therefore special care should be taken in re-chucking. Whenever possible, toleranced features and datum features related by tolerances of orientation or location should be manufactured without re-chucking.

15.2 Recommendations for manufacturing

In order to respect **geometrical tolerances**, the abilities of the machine tools should be assessed and classified. In order to respect narrow geometrical tolerances, by the

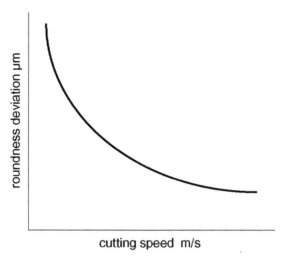

Fig. 15.3 Roundness deviation as a function of cutting speed

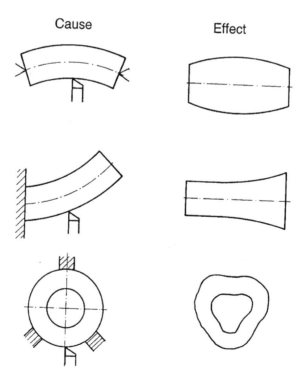

Fig. 15.4 Influence of chucking and cutting pressure on the workpiece form

classification, the process data (e.g. cutting speed) and workpiece properties (e.g. wall thickness) must be taken into account. Because re-chucking influences the geometrical deviations greatly, it should be specified in the manufacturing planning whether it is to be re-chucked, and if so then, when, where, how and with what precision.

With regard to respecting the **general geometrical tolerances** see 16.

In order to respect the **envelope requirement** Ⓔ, it is recommended that the tolerance specified in the drawing at the maximum material limit be reduced by the expected form deviation, i.e. the actual sizes must differ from the maximum material size at least by the amount of the expected form deviation.*

In order to respect the **maximum material requirement Ⓜ at the toleranced feature**, the following are recommended:

- withdrawing indication 0 Ⓜ: the same applies as for Ⓔ;*†
- withdrawing indication t Ⓜ (where $t > 0$). When this tolerance t is larger than the expected geometrical deviation, it may be manufactured as if Ⓜ were not specified. Only when the specified geometrical tolerance is exceeded is it necessary to check according to the maximum material requirement.† When the indicated tolerance t is not larger than the expected geometrical deviation, it is recommended to reduce the size tolerance at the maximum material limit accordingly.*†

In order to respect the **maximum material requirement Ⓜ at the datum feature**, the following is recommended:

- when at the datum letter in the tolerance frame the symbol Ⓜ is indicated but the datum feature itself has no geometrical tolerance with specified Ⓜ: the same applies as for Ⓔ;*†
- when at the datum feature itself a geometrical tolerance with specified Ⓜ is indicated, proceed in the same way as with t Ⓜ (where $t > 0$) at toleranced features, as given above.*†

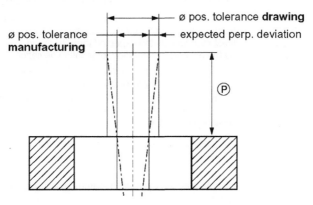

Fig. 15.5 Recommendation for manufacturing in order to respect the projected tolerance zone Ⓟ

---

* If appropriate, a manufacturing drawing should be issued using the reciprocity requirement (see 9.3.6 and Figs 20.62 ff).

† Features related by relatively small tolerances of orientation or location (toleranced feature(s) and datum feature(s)) should be manufactured without re-chucking.

In order to respect the **least material requirement** Ⓛ, proceed in the same way as with the maximum material requirement, but reduce the size tolerance at the least material limit (see 11.3 and 11.4).

In order to respect the **projected tolerance zone** Ⓟ, it is recommended that the positional tolerance be reduced by twice the expected angularity deviation related to the projected length (Fig. 15.5).

In order to respect the **reciprocity requirement** Ⓡ, the manufacturing may be as if Ⓡ were not specified.

Figure 15.6 gives a synopsis of these recommendations. However, for the decision as to whether the workpiece meets the drawing specifications the definitions of the requirements must be observed.

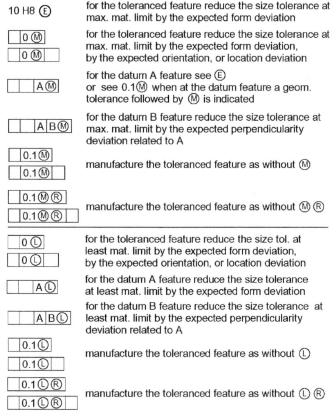

Features related by relatively small tolerances of orientation or location (toleranced feature(s) and datum feature(s)) should be manufactured without re-chucking.

Fig. 15.6 Recommendations for manufacturing in order to respect Ⓔ, Ⓜ, Ⓛ and Ⓡ

16

General Geometrical Tolerances

16.1 Demand for general geometrical tolerances

As the principle of independency demands an indication for each requirement, it calls for the application of a standard on general geometrical tolerances (title block tolerances on geometry). Otherwise, the drawing would be embroidered with geometrical tolerance indications (Fig. 16.11).

Even if Rule #1 of ASME Y14.5M (see 21.1.2) is used, there is still a need for general geometrical tolerances on orientation (perpendicularity) and location (coaxiality and symmetry), because Rule #1 does not apply to related features. Furthermore, there is a need for general geometrical tolerances on form for single features to which a disclaimer from Rule #1 is indicated.

16.2 Concept of general tolerances

Before applying general tolerances, their concept must be agreed upon. This concept has been developed in the International Organization for Standardization (ISO), as follows (ISO 2768):

(a) Each feature requires limits for its deviations determined by its function.
(b) The drawing must be definitive, i.e. the drawing must specify all dimensional and geometrical tolerances necessary to completely define the shape and size of the part.
(c) Above certain values of tolerances, there is generally no gain in economy by increasing them further. These tolerances are not exceeded in normal workshop practice without particular effort. This is the normal (customary) workshop accuracy.
(d) General tolerances take account of the normal workshop accuracy (Fig. 16.1). The general tolerances are specified on the drawing by a general tolerance note, usually in or near the title block, e.g. referring to a standard specifying the general tolerances.
(e) If the function requires tolerances smaller than the general tolerances, the required tolerances must be indicated individually.

   If the function allows tolerances equal to or larger than the general tolerances, they should not be indicated individually. For the features concerned the general tolerances should apply.
(f) Exceptions to the rule are where the functions allow larger tolerances and these larger tolerances are more economical in the particular cases than the general tolerances (e.g. the length of blind holes drilled at assembly). In these few exceptional

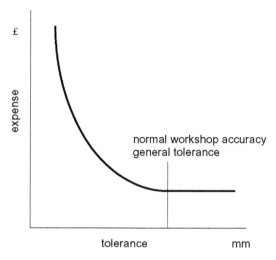

Fig. 16.1 Cost function of tolerances; general tolerances

cases the tolerance required by the function, although larger than the general tolerance, should be indicated individually.

(g) Before the introduction of general tolerances into the workshop, the tolerance class (accuracy grade) of general tolerances to which its normal workshop accuracy corresponds should be assessed. (The general tolerances of that tolerance class should be reliably respected, i.e. they should be so large that they are well within the manufacturing capability of the ordinary machine tools, see, e.g., Fig. 16.2.)

(h) The workshop should only accept drawings where the general tolerance class is equal to or coarser than its normal workshop accuracy.

(i) The workshop should ensure by sampling inspection that its normal workshop accuracy is not impaired in the course of time.

(j) If a general tolerance is exceeded (although this is very unlikely, see (c), (g), (h), and (i)), the decision whether or not the part is to be rejected depends on the function (see (c) and (e)).

General tolerances are not guidelines that may be exceeded without risk, as it has sometimes been wrongly supposed. The probability that exceeding them will lead to rejection of the part is merely less than with individually indicated tolerances.

(k) The use of general tolerances leads to the following advantages:
- drawings are easier to read (not being embroidered with tolerance indications) (Fig. 16.11);
- the designer saves a great deal of detailed tolerance calculations (it is sufficient to know that the function allows tolerances = general tolerances);
- drawings indicate which characteristics can be produced with normal manufacturing capability (ordinary machine tools);
- drawings indicate which characteristics are not likely to be exceeded (checking level can be less frequent).

Sometimes concepts are proposed that deviate from the above, for example for distinguishing between functional dimensions and non-functional dimensions. The tolerances for functional dimensions should then be indicated individually, whereas the general tolerances should apply only to the non-functional dimensions. As mentioned above, there are limits derived from the function for each dimension (feature). Therefore each dimension is a functional one, and it is impossible to distinguish between them under these circumstances. A similar argument applies to geometrical characteristics.

Other concepts propose applying the general tolerances only to those characteristics the functional tolerance for which is much larger than the general tolerance. But then the question arises as to by how much the functional tolerance should be larger than the general tolerance in order to apply the latter. If a factor is established (e.g. 2) then why not double the general tolerance?

An objection sometimes heard against the general geometrical tolerances is that they would also control characteristics that are not essential for the function of the workpiece (see Fig. 20.18). Since the general tolerances are so large that they are respected in any case (see 16.2(c)), there is no disadvantage in applying them to features that would allow wider tolerances.

What logically remains is the ISO concept given above under (a) to (k).

16.3 Derivation and application of general geometrical tolerances

The general geometrical tolerances for machined parts that are now standardized in ISO 2768-2 and that have been in use within parts of the industry for many years (e.g. in Germany) are described in the following.

These general geometrical tolerances, given in Tables 16.1 to 16.4, are derived from measurements on workpieces formed by metal removal. Mainly workpieces were measured to which the general dimensional tolerances ISO 2768-m had been applied (manufacturing of turbines, machine tools, fine mechanical engineering, etc.). Only such features were measured that had no geometrical tolerance indication on the drawing.

16.3.1 Straightness and flatness

The general tolerances on straightness and flatness are given in Table 16.1. When selecting a straightness tolerance from the table, the length of the corresponding line must be

Table 16.1 General tolerances on straightness and flatness (values in mm)

Accuracy grade	Straightness and flatness tolerances for ranges of nominal length					
	to 10	over 10 to 30	over 30 to 100	over 100 to 300	over 300 to 1000	over 1000 to 3000
H	0.02	0.05	0.1	0.2	0.3	0.4
K	0.05	0.1	0.2	0.4	0.6	0.8
L	0.1	0.2	0.4	0.8	1.2	1.6

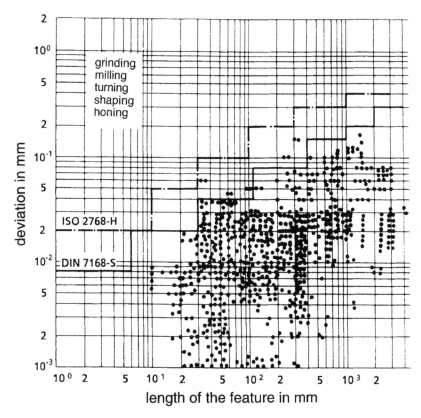

Fig. 16.2 Measured straightness deviations (measured in German and Japanese industry)

taken, and in the case of a flatness tolerance the longer length of the surface or the diameter of the circular surface must be taken.

The general tolerances given in Table 16.1 are derived from measurements as described above (Fig. 16.2). The measured geometrical deviations have remained within the tolerances of class H in Table 1. This has been proved by measurements made in the industry of different countries (e.g. in Germany, Japan and the UK).

The general tolerances of classes K and L are multiples of the tolerances of class H and may be applied when necessary.

16.3.2 Roundness (circularity)

General tolerances on circularity have been established equal to the numerical value of the diameter tolerance (Fig. 16.3), or to the respective value of the general tolerance on circular run-out in Table 16.4, whichever is the smaller.

The circularity deviation cannot be larger than the radial circular run-out deviation of the same feature (for geometrical reasons). Therefore the numerical values of the

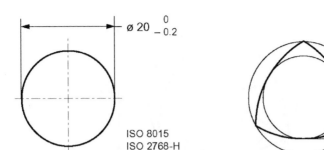

Fig. 16.3 Roundness tolerance zone, drawing indication and interpretation

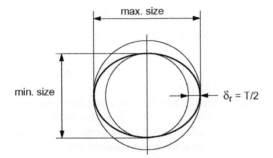

Fig. 16.4 Maximum permissible elliptic form deviation

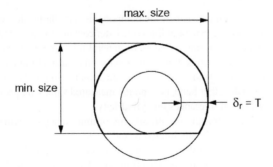

Fig. 16.5 Maximum permissible form deviation

general tolerances on circular run-out (Table 16.4) have been taken as the upper limits on the general tolerances on circularity.

Whether the roundness deviation may take advantage of the full tolerance zone depends on the diameter tolerance and on the shape of the roundness deviation.

In the case of an elliptic form the deviation may only occur up to half of the numerical value of the size tolerance, otherwise actual local sizes would exceed the size tolerance (Fig. 16.4).

In the case of Fig. 16.5 the form deviation may occur up to the size tolerance.

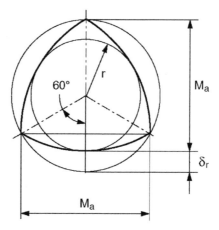

Fig. 16.6 Maximum possible form deviation with three-lobed form

In the case of a three-lobed form the deviation can occur up to 0.15 times the maximum permissible diameter M_a (Fig. 16.6).

This might be more than the size tolerance, and could then be more than what is allowed by the general tolerance on circularity. Although such extreme form deviations seldom occur, those production methods in which lobed forms can appear should be observed in order not to exceed the general tolerances on circularity.

16.3.3 Cylindricity

The cylindricity deviation consists of the components: circularity deviation and parallelism deviation of opposite generator lines (the latter contains the straightness deviation). Each of these components is controlled by its general tolerance. However, it is unlikely that on one workpiece the two components occur with their maximum permissible values and that they accumulate to the theoretical maximum permissible value of the cylindricity deviation. On the other hand, not enough measured values are presently available to derive suitable general tolerances on cylindricity.

Since the cylindricity deviation is almost only of importance for cylindrical fits, and since the form deviation is already limited by the envelope requirement Ⓔ or by an individually indicated circularity tolerance, there seems to be no need for the establishment of general tolerances on cylindricity. Therefore they have been omitted.

16.3.4 Line profile, surface profile and position

General tolerances for line profile and surface profile are not standardized. These characteristics have to be toleranced individually (see, e.g., Table 6.1) or by general tolerances according to a company standard.

16.3.5 Parallelism

The limitation for parallelism deviations results from the straightness (or flatness) tolerance (Fig. 16.7(a)), or from the size tolerance (Fig. 16.7(b)), whichever is the larger.

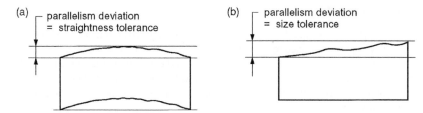

Fig. 16.7 Parallelism deviation: (a) equal straightness tolerance; (b) equal size tolerance

Larger parallelism deviations are not to be expected, because otherwise either the straightness or flatness tolerance or the size tolerance would be exceeded.

16.3.6 Perpendicularity

The general tolerances on perpendicularity are given in Table 16.2. The longer of the two sides forming the rectangular angle is to be taken as the datum. If the nominal lengths of the two sides are equal, either of them may apply as the datum.

In deriving the values of Table 16.2 the following has been observed.

The definition of the geometric tolerance on perpendicularity according to ISO 1101 is as follows:

Zone between two parallel straight lines or planes or cylindrical zone that is perpendicular to the datum and within which the actual line or surface or the axis or median plane shall remain.

This tolerance zone also limits the straightness or flatness deviations and the axial run-out deviations of the sides forming the rectangular angle. Therefore the general tolerance on perpendicularity should not be smaller than the general tolerances on straightness (and flatness) and on axial run-out.

The general tolerances on perpendicularity given in Table 16.2 respect this, and are derived from measurements as described above. The measured deviations have remained within the tolerances of class H in Table 16.2.

Table 16.2 General tolerances on perpendicularity (values in mm)

Accuracy grade	Perpendicularity tolerances for ranges of nominal length of the shorter side			
	to 100	over 100 to 300	over 300 to 1000	over 1000 to 3000
H	0.2	0.3	0.4	0.5
K	0.4	0.6	0.8	1
L	0.6	1	1.5	2

16.3.7 Angularity

General tolerances on angularity in the definition of ISO 1101 (zone between two parallel lines or planes inclined to the datum in the theoretically exact angle indicated in a rectangular frame) are not specified. For angles indicated in angular dimensions the general tolerances according to ISO 2768-1 apply. For angles indicated in theoretically exact dimensions (e.g. 30°) the angularity tolerance has to be indicated on the drawing individually.

16.3.8 Coaxiality

In the extreme case the coaxiality deviation may be as great as the value given in Table 16.4 for the radial circular run-out tolerance, since the radial circular run-out deviation is composed of the coaxiality deviation and parts of the circularity deviation. General tolerances on coaxiality are not intended to be standardized.

16.3.9 Symmetry

The general tolerances on symmetry are given in Table 16.3. They apply to symmetrical features, also if one of the two features is symmetrical and the other cylindrical. The longer feature is to be taken as the datum. If the nominal lengths of the two features are equal, either of them may apply as the datum.

In deriving the values of Table 16.3 the following has been observed.

As the tolerance zone on symmetry also limits certain straightness or flatness deviations, the general tolerances on symmetry should not be smaller than the general tolerances on straightness and flatness.

Furthermore, measurements on workpieces (as described above) revealed that symmetry deviations up to 0.5 mm occur independently of the feature length (Fig. 16.8).

The reason for this is probably the following. The general tolerances are set up depending on the largest measured deviations. These deviations were not due to the inaccuracy of the machine tool but rather to inaccuracy when adjusting the workpiece in the machine tool after the workpiece has been turned over (re-chucking). Small and large workpieces were adjusted with the same inaccuracy, and showed the same distribution of the measured symmetry deviations.

Table 16.3 General tolerances on symmetry (values in mm)

Accuracy grade	Symmetry tolerances for ranges of nominal lengths			
	to 100	over 100 to 300	over 300 to 1000	over 1000 to 3000
H	0.5			
K	0.6		0.8	1
L	0.6	1	1.5	2

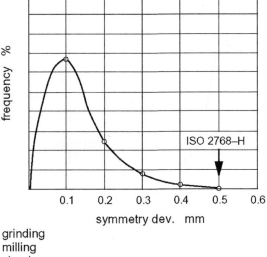

grinding
milling
shaping

length of the features up to 500 mm

Fig. 16.8 Measured symmetry deviations

16.3.10 Circular run-out

The general tolerances on circular run-out* (radial, axial and inclined circular run-out) are given in Table 16.4. The bearing surfaces are to be taken as the datum if they are designated as bearing surfaces. In the other case, for radial circular run-out, the longer feature is to be taken as the datum. If the nominal lengths of the two features are equal, either of them may apply as the datum.

Table 16.4 General toler-ances on circular run-out (values in mm)

Tolerance class	Run-out tolerance
H	0.1
K	0.2
L	0.5

* Circular run-out tolerances according to ISO 1101 do not limit the deviations of the straightness of cylinder generator lines or the flatness of faces, which is in contrast to total run-out tolerances.

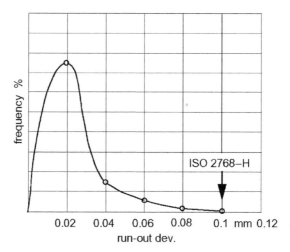

Fig. 16.9 Measured circular radial run-out deviations

Table 16.4 is derived from measurements on workpieces as described above. The measured geometric deviations have remained within the tolerances of class H in Table 16.4 (Fig. 16.9).

The measured deviations have not shown any relationship to

- diameter or length (assessed up to diameter 900 mm);
- size tolerance;
- specified surface roughness;
- material (assessed metallic material only).

This has been proved during use since 1974 in, for example, the German industry.

The reason for this is probably the following. The general tolerances were set up depending on the largest measured deviations. These deviations were not due to the inaccuracy of the machine tool but rather to the inaccuracy when adjusting the workpiece in the machine tool after the workpiece has been turned over (re-chucking). During adjustment, the workpiece was usually measured where the largest deviations are to be detected (e.g. at the outer diameter). Therefore small and large workpieces were adjusted with the same inaccuracy.

16.3.11 Total run-out

The radial total run-out deviation consists of the components: circular run-out deviation and parallelism deviation (the latter contains the straightness deviation). The axial total run-out deviation consists of the components: circular run-out deviation and flatness deviation. Each of these components is controlled by its general tolerance.

General tolerances on radial total run-out are not intended to be standardized, for similar reasons as for cylindricity. General tolerances on axial total run-out are not intended to be standardized, since general tolerances on perpendicularity are already standardized.

16.4 Datums

For general tolerances of orientation, location and run-out it is necessary to determine the datums without drawing indications. According to ISO 2768-2, the longer of the two considered features applies as the datum. When they are of equal nominal length, either may serve as a datum.

An exception applies with general tolerances of run-out when there are bearing surfaces designated as such in the drawing. Then these surfaces serve as the datum(s).

Although, with the exception of designated bearing surfaces, datums for general geometrical tolerances are not designated in the drawing, there is no accumulation of general geometrical tolerances possible from one feature to the next etc., because the general geometrical tolerances apply to all possible combinations of any two features of the workpiece.

It may occur that the two features in a combination for related general tolerances (e.g. run-out) are short in relation to their diameter. Therefore the datum is not inspection appropriate and individual tolerancing should be used, see, e.g., Fig. 19.14.

16.5 Indication on drawings

When the general tolerances according to ISO 2768-2 shall apply, this has to be designated in or near the title block, e.g.:

General Tolerances   ISO 2768-mH

In this example m stands for general dimensional tolerances class m and H for general geometrical tolerances class H.

The indication on drawings when the envelope requirement (Rule #1 of ASME Y14.5M or DIN 7167) is used in addition to the general tolerances as a general requirement is dealt with in 16.6.

16.6 Envelope requirement in addition to general form tolerances

It was a strong desire from some countries, at least for a transition period, to allow in addition to the general tolerances the use of the envelope requirement (Rule #1 of ASME Y14.5M or DIN 7167) without individual indications for all single features of size. (Single features of size consist of a cylindrical surface or two parallel opposite plane surfaces.) This means limiting form deviations also by the envelope requirement even when the feature has not the function of a fit and the feature's function would allow certain form deviations when it is everywhere at maximum material size.

Drawing indication:　　　　Interpretation:

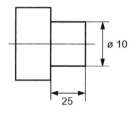

(1) actual local sizes (two point measurements) between ⌀ 9.8 and ⌀ 10.2 (ISO 2768-m)

(2) straightness deviations ≤ 0.05 (ISO 2768-H)

(3) envelope requirement to be respected (ISO 2768-E)

(4) deviations of orientation and location according to ISO 2768-H (e.g. circular run-out deviations ≤ 0.1)

ISO 8015
ISO 2768-mH-E

Fig. 16.10 Envelope requirement as general requirement, example according to ISO 2768-2

In these cases the drawing shall be designated in or near the title block, e.g.:

General Tolerances　　ISO 2768-mH-E

With this designation, the envelope requirement applies to all individual features of size, provided that there is no individually indicated form tolerance applied that is larger than the size tolerance. The envelope requirement is considered as an additional requirement. All the other indications (dimensional and geometrical tolerances) retain their meaning as described above (Fig. 16.10). See also 17.3.

This concept is sometimes referred to as the principle of dependency.

16.7　Application of the general geometrical tolerances according to ISO 2768-2

When the general geometrical tolerances according to ISO 2768-2 shall apply, this has to be stated on the drawing (see 16.5).

The general tolerances according to ISO 2768-2 are applicable both when the principle of independency applies and when the principle of dependency applies (see 16.6 and 17).

When the principle of dependency shall apply, see 16.6.

For the application of the general geometrical tolerances according to ISO 2769-2 the normal workshop accuracy should be known. The general geometrical tolerances should be not smaller than the normal workshop accuracy. The normal workshop accuracy can be assessed by measurements of workpieces manufactured under normal workshop (production) conditions. The normal workshop accuracy (regarding geometrical deviations) depends on the accuracy of the (most inaccurate) machine tools and on the accuracies of adjustment of the workpieces during re-chucking on the machine tool.

Normally the normal workshop accuracy in mechanical engineering corresponds to ISO 2768-H (and to ISO 2768-mH when general tolerances for dimensions are included).

Table 16.5 General geometrical tolerances for machining according to ISO 2768-2 synoptic table

Characteristic	General tolerance
−	Table 16.1
⊡	Table 16.1
○	Size tolerance or Table 16.4*
∥	Size tolerance or Table 16.1†
⊥	Table 16.2
≡	Table 16.3
◢	Table 16.4
◎	See ◢
⌀	See ○ + ∥
↗	Radial run-out see ◢ + ∥
	Axial run-out see ⊥ or ◢ + ⊡ *
⌒	
⌓	Not specified
∠	
⊕	

*Whichever is the smaller.
†Whichever is the larger.

An alternative to the general tolerances according to ISO 2768 is the use of a general tolerance on surface profile (with or without a datum system) according to a company standard. For holes a general positional tolerance (with a datum system and with the maximum material requirement) may be specified in addition.

Usually, if not otherwise specified, the general positional tolerance overrides the general surface profile tolerance.

Important is that the general tolerances are so large that they will be respected with a very high probability, so that manufacturing and inspection can concentrate on individually indicated tolerances (see 16.2).

16.8 General tolerances for castings

International standards on general geometrical tolerances for castings have not yet been published, but are in preparation (ISO 8062-3). Measurements on castings in the industry of various countries have been made and the results submitted to the committee ISO/TC 213 in order to derive an appropriate tolerance system for general tolerances.

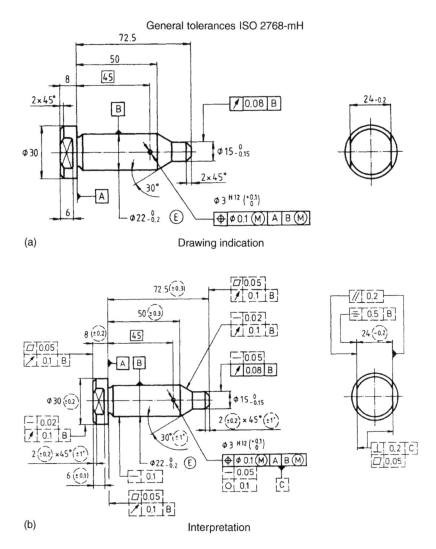

(a) Drawing indication

(b) Interpretation

Tolerances shown in dashed lines (boxes and circles) are general tolerances according to ISO 2768-mH which will be automatically respected by machining in a workshop with a customary accuracy equal to or finer than ISO 2768-mH and would not normally require to be inspected.

Fig. 16.11 Example of applying general tolerances

16.9 General tolerances for welded parts

General geometrical tolerances for welded parts are standardized in ISO 13 920-2. The standard specifies only general tolerances on straightness, flatness and parallelism for features produced by welding. Table 16.6 gives these tolerances. They apply under the principle of independency ISO 8015 (see 17).

Table 16.6 General tolerances of straightness, flatness and parallelism for welded parts according to ISO 13920-2

Tolerance grade	Nominal length of the longer side in mm									
	above 30 to 120	above 120 to 400	above 400 to 1000	above 1000 to 2000	above 2000 to 4000	above 4000 to 8000	above 8000 to 12 000	above 12 000 to 16 000	above 16 000 to 20 000	above 20 000
	Tolerance in mm									
E	0.5	1	1.5	2	3	4	5	6	7	8
F	1	1.5	3	4.5	6	8	10	12	14	16
G	1.5	3	5,5	9	11	16	20	22	25	25
H	2.5	5	9	14	18	26	32	36	40	40

Tolerances on perpendicularity, angularity, coaxiality and symmetry are to be indicated individually.

16.10   General tolerances for parts out of plastics

An international standard on general geometrical tolerances for parts out of plastics does not exist. A company standard should be established in order to use the advantages as described in 16.2(k).

17
Tolerancing Principles

17.1 Limitation by function

The function of a feature limits the permissible deviations of size, form, orientation and location. If one of these deviations is too large it affects the functioning of the feature. This applies to each feature and may be illustrated by the following example of a flange shown in Fig. 17.1.

The deviations allowed by the function may sometimes be relatively large, e.g. for D in Fig. 17.1, but there are always limits given by the function. Otherwise the considered feature would not need to exist.

17.2 Need for completely toleranced shape

Consequently the tolerance for each characteristic should be given. As there is hardly any communication other than by technical drawings the reliance on undefined workshop accuracy or good workmanship is not sufficient. The international exchange of technical

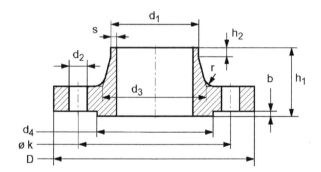

D	Reduction limited by the distance to the hole d_2
	Increase limited by the adjoining parts
d_2	Reduction limited by the bolt
	Increase limited by the pressure under the bolt
r	Reduction limited by the fatigue notch effect
	Increase limited by the bolt face
d_3	Reduction limited by the strength
	Increase limited by the bolt face

etc.

Fig. 17.1 Flange

drawings requires that the drawings specify all dimensional and geometrical tolerances necessary to define the shape and size of the part completely.

17.3   Situation in the past

Regarding geometrical tolerances this requirement has not always been respected. In the past, in most countries there has been a reliance on undefined good workmanship regarding deviations of perpendicularity, coaxiality and symmetry for features that have no corresponding tolerance indication on the drawing.

In some countries – contrary to others – the form deviations have been assumed to be controlled by the dimensional (size) tolerances. This is specified, for example, by Rule #1 of the USA Standard ASME Y14.5M-1994.

This requires that the feature (cylindrical surface or two parallel opposite plane surfaces) shall be contained within an envelope of perfect form at its maximum material size (maximum size of a shaft or minimum size of a hole). In other words a shaft at maximum material size everywhere shall have neither straightness nor circularity deviations, i.e. it shall be of perfect form.

This restriction is necessary for fits but in most other functional cases it is an unnecessary restriction. Reviews of technical drawings in different countries revealed that generally less than 10% of the features (dimensions) on a drawing should be associated with the Rule #1. More than 90% of the features do not need this restriction. Their function allows geometrical deviations even when the feature is at its maximum material size everywhere.

To features for which Rule #1 is not required a disclaimer should be indicated on the drawing. But this has usually not been done because it would have embroidered the drawing with disclaimers. On the other hand the drawing should call out what is required, rather than what is not required. (When disclaimers are indicated and a disclaimer is missing it might be doubtful whether Rule #1 is in fact required or the draftsman has simply overlooked this dimension among the many disclaimers to be indicated in the drawing.)

In the inspection of workpieces the requirement of Rule #1 has been checked only if a full form gauge was used. However, in most cases snap gauges or measuring instruments have been used, and the requirements of Rule #1 have not been verified (Fig. 17.2).

In order to respect Rule #1 the feature must be manufactured with actual local sizes that depart from the maximum material size at least by the amount of the expected straightness and certain circularity deviations. In other words it is not allowed to take full advantage of the dimensional tolerance for the actual (local) sizes. In frequent cases this has not been respected.

From measured workpieces it is known that the magnitude of form deviations permitted by Rule #1 is not in line with the normal workshop accuracy of various industrial companies. The form deviations that occur through normal machining processes are sometimes greater than small dimensional tolerances. About 10 000 measurements obtained in 17 companies in Switzerland revealed that in about 20% the straightness deviation was larger than the dimensional tolerance. As there are also roundness deviations and variations in size to be taken into account much more than 20% violate Rule #1, although the companies pretended to manufacture according to Rule #1. From measurements in Germany, Japan and UK the same derives. On the other hand large dimensional tolerances, e.g. h14, are much larger than the form deviations that occur through

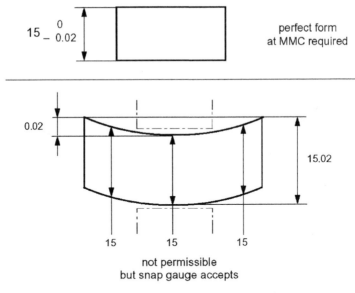

$15 \, {}^{\,0}_{-\,0.02}$ perfect form
at MMC required

0.02

15.02

15   15   15

not permissible
but snap gauge accepts

Fig. 17.2 Limitation of straightness deviations by size tolerance

normal machining processes (compare with Table 16.1). Therefore general form tolerances, as described below, would be more in line with the normal manufacturing capability than reliance on Rule #1.

This has led to the fact that the relationship between dimensional and geometrical tolerances has been described and defined in different ways in various countries. Some countries have adopted Rule #1; others have treated dimensional and geometrical tolerances independently from each other. Some countries have regarded an indicated geometrical tolerance to a feature as a disclaimer of Rule #1; other countries have not. In some countries it has been assumed that the feature shall not violate the boundary of perfect form at least material size; in others it has not. Some countries have regarded a dimensional tolerance adjacent to a positional toleranced pattern different from a dimensional tolerance adjacent to a dimensional tolerance; other countries have not.

Even within companies opinions have varied. Generally while the designers have tended towards the thinking of Rule #1, the inspectors have referred to the fact that generally size and form (and orientation and location) are inspected independently from each other.

17.4 Principle of independency

As this situation was considered unsatisfactory, the relevant ISO Committees decided to standardize in ISO 8015 the principle of independency as the fundamental tolerancing principle. It states:

Each requirement for dimensional or geometrical tolerances specified on a drawing shall be met independently, unless a particular relationship is specified, i.e. Ⓜ or Ⓔ or Ⓛ.

Drawing indication:

Interpretation:

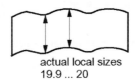

actual local sizes
19.9 ... 20

Fig. 17.3 Principle of independency; linear dimensional (size) tolerance

This means linear dimensional tolerances (e.g. ± 0.1 or H7) control only actual local sizes (two-point measurements), e.g. wall thicknesses, but not the form deviations of the feature, Fig. 17.3.

Angular dimensional tolerances, specified in angular units (e.g. $\pm 1°$), control the general orientation of lines or line elements (section lines) of surfaces but not their form deviations. The general orientation of the line derived from the actual surface is the orientation of the contacting line of perfect form (contacting straight line). The maximum distance between the contacting line and the actual line should be the least possible (Fig. 3.45).

When no relationship is specified the geometrical tolerance applies regardless of feature size and the two characteristics are treated (inspected) as unrelated requirements (Figs 3.45, 17.3 to 17.5).

Consequently, the principle of independency demands a separate indication for each requirement (which may cause a particular checking operation and may be inspected regardless of other characteristics). Therefore **it is sufficient to inspect what is indicated in the drawing**. The requirements are indicated in an analytical way. The requirements may be:

- linear dimensional tolerance (ISO 8015, ISO 286);
- angular dimensional tolerance (ISO 8015);
- form tolerance (ISO 1101);
- tolerance of orientation, location, run-out (ISO 1101);
- envelope requirement Ⓔ (ISO 8015, ISO 286);
- maximum material requirement Ⓜ (ISO 2692);
- least material requirement Ⓛ (ISO 2692);
- reciprocity requirement Ⓡ (ISO 2692);
- projected tolerance zone Ⓟ (ISO 10 578);
- tolerances in the restrained condition ISO 10 579, tolerances in the free state Ⓕ.

There are no "hidden rules", as, e.g., Rule #1, i.e. there are no rules that are not expressively indicated on the drawing.

Drawing indication:

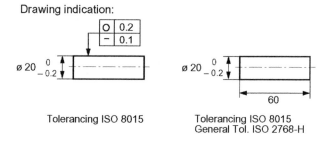

Tolerancing ISO 8015

Tolerancing ISO 8015
General Tol. ISO 2768-H

Interpretation, extreme permissible part:

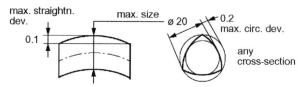

Fig. 17.4 Principle of independency; size and form

Drawing indication:

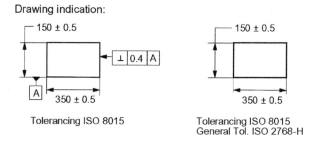

Tolerancing ISO 8015

Tolerancing ISO 8015
General Tol. ISO 2768-H

Interpretation, extreme permissible part:

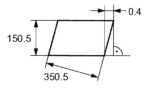

Fig. 17.5 Principle of independency; size and orientation

According to the principle of independency, form deviations are not controlled by the size tolerances, hence it calls for the application of general tolerances of form. However, the application of general geometrical tolerances is necessary anyway in order to control deviations of orientation and location (see 16).

Drawings to which general geometrical tolerances are not applied and which do not specify all necessary geometrical tolerances are incomplete. They may cause difficulties

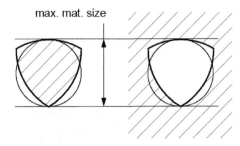

max. mat. size

Fig. 17.6 Violation of the maximum material size circle by a lobed form

when the designer has assumed more precise normal workshop accuracy than the manufacturer has implied.

The application of the principle of independency (ISO 8015) together with general geometrical tolerances (e.g. ISO 2768) and the symbol Ⓔ, where applicable, provide the following advantages:

- the same interpretation of drawings worldwide (no misunderstandings due to different national rules);
- drawings can be toleranced according to the functional requirements (no unnecessary limitations like, e.g., Rule #1);
- drawing indications show what is required and not what is not required (no disclaimer Ⓕ necessary);
- drawings are decisive (fully toleranced);
- tolerancing in accordance with inspection practices (analytical method, each check requires a separate requirement indication, no hidden rules);
- drawings are more clear (drawings are not embroidered with tolerance indications);
- there is a foundation for economical production.

General geometrical tolerances together with the symbols Ⓜ, Ⓛ, Ⓡ, Ⓟ and Ⓕ, where applicable, should be applied in any case (also when the principle of dependency is applied; see 17.6).

It should be noted that in cases of extreme lobed form deviations the feature cross-section can violate the circle of the maximum material size (Fig. 17.6). The extreme permissible part with maximum material size is shown in Fig. 17.4. When this is not allowed, e.g. in the case of a fit, the envelope requirement is to be applied (see 10).

The maximum perfect cylinder that is contained in the smallest permissible cylindrical part (shaft) is smaller than the least material size by the amount of the sum of the tolerance on circularity and the tolerance on straightness because of the lobed and bent form deviation (Fig. 17.7). This has to be taken into account, e.g. with stock material.

17.5 Identification of drawings

Drawings to which the principle of independency applies have to be distinguished from those to which other (e.g. former) rules apply and are therefore to be identified according to ISO 8015 by the title box indication: Tolerancing ISO 8015.

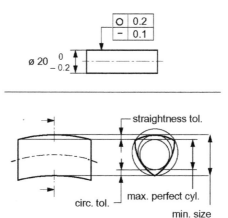

Fig. 17.7 Maximum perfect cylinder (shaft) contained in the least part

17.6 Principle of dependency

Some national standards provide a tolerancing concept, the principle of dependency, as an alternative concept to the principle of independency according to ISO 8015.

The principle of dependency specifies for features of size (i.e. established by cylinders or two opposite parallel planes) the envelope requirement without any drawing indication, e.g. ASME Y14.5M, Rule #1, former BS 308, DIN 7167 (see 21).

This applies to size and form of isolated features only. The principle of dependency does not apply to orientation or location of related features. For example, a cube consists of three individual features of size each composed of a set of two opposite parallel plane surfaces. The perpendicularity of those individual features of size is not controlled by their size tolerances. It is controlled by general geometrical tolerances, if applied, or by individually indicated perpendicularity tolerances (see, e.g., Fig. 20.63).

The principle of dependency does not apply to angular dimensions and angular dimensional tolerances, as, e.g., ±1°. Most national standards, e.g. BS 308, define angular dimensional tolerances as limits for contacting lines, as defined in ISO 8015 (Fig. 3.45), i.e. straightness and flatness deviations are not controlled by the angular tolerance. However, ASME Y14.5M defines angular tolerances differently (see 21.1.2).

The principle of dependency can be indicated on drawings, when general geometrical tolerances according to ISO 2768 are applied, by the letter "E" in the title box identification, e.g. ISO 2768-mH-E (see 16.6).

17.7 Choice of tolerancing principle

For new or altered drawings it has to be decided whether the principle of independency (ISO 8015) or the principle of dependency (ISO 2768- … -E, or national standard, e.g. ASME Y14.5M Rule #1, DIN 7167) shall be applied (see also 21).

The principle of dependency has the advantage that the designer does not need to check where the envelope requirement is necessary. In this respect the principle of dependency is failsafe.

	Principle of independency ISO 8015	Principle of dependency ASME Y14.5; DIN 7167
design	Ⓔ or Ⓜ to be indicated (thinking necessary)	failsafe (no thinking necessary)
manufacturing	diminishing of tolerance	
	with Ⓔ only	always*
inspection	check of envelope requirement	
	with Ⓔ only	always*

*or unnecessary small dimensional tolerance.

Fig. 17.8 Comparison of the principle of independency ISO 8015 with the principle of dependency

Besides, with the principle of dependency the following has to be observed:

- To respect the envelope requirement the manufacturer must diminish the tolerance at the maximum material side by the amount of the expected form deviations. According to the drawing this applies to all features of size, even when there is no functional necessity.
- With the inspection it is not sufficient to inspect the dimensional deviations and the form deviations. In addition it has to be checked whether the envelope requirement has been respected (see 10 and 18.7.11).
- The principle of dependency tempts the designer to specify narrow dimensional tolerances although wider dimensional tolerances and (narrower) geometrical tolerances would be sufficient.
- Because it is well known that the envelope requirement with the principle of dependency is often violated there is the danger that the designer might specify narrower dimensional tolerances than necessary according to the function.

These all lead to unnecessarily great effort (and cost) for manufacturing and inspection or to a large number of workpieces that do not comply with the drawing.

In contrast, the principle of independency ISO 8015 requires slightly more effort in the design, but enables function-related designs with all advantages in manufacturing and inspection. The envelope requirement must be respected only where it is a functional need. This is a prerequisite for economical production taking into account the requirements of the quality assurance according to ISO 9001 etc. Figure 17.8 gives a comparison of the two principles.

Regarding the application of general geometrical tolerances (e.g. according to ISO 2768) and of Ⓜ, Ⓛ, Ⓡ, Ⓟ and Ⓕ, there is no difference between both principles.

18

Inspection of Geometrical Deviations

18.1 General

The following deals with generalities on the inspection of geometrical deviations. A synopsis on inspection methods is given in ISO TR 5460. More detailed descriptions of inspection methods are given in the former East German Standards TGL 39 092 to TGL 39 098 and TGL 43041 to 43045, TGL 43 529 and TGL 43 530.

Geometrical tolerances are geometrically exactly defined. They determine geometrical zones within which the surface of the feature must be contained.

There are several methods for inspecting whether the geometrical tolerances have been respected. These methods are more or less precise. The drawing indications according to ISO 1101 do not prescribe a particular (certain) inspection method. According to ISO TR 14 253-2 the PUMA method should be used in order to optimize manufacturing and inspection (see 18.11).

In industrial practice the inspection of geometrical deviations is often only economical with less precise inspection methods. Less precise inspection methods are normally less time consuming and less costly. In contrast, more precise inspection methods are normally more time consuming and more costly. Therefore it is often advisable to start the inspection with a cheap but less precise method and to switch to a more precise (and more expensive) method only in cases where the measurement result is near the limit given by the tolerance (see, e.g., 18.7.9.3.6).

A prerequisite is that the possible errors of the inspection methods are known. Information on systematic errors of inspection methods is given in ISO TR 14 253-2 (see 18.11) and in the East German Standards mentioned above.

Specifications on the necessary sample sizes (number of traces, number of probed points) are not internationally standardized. They depend on the size of the feature to be inspected and on the ratio between form deviation and geometrical tolerance. The measurement result shall be representative of the feature. Some hints are given in 18.9.

Particular inspection methods should not be prescribed by the drawing indication for the following reasons.

- The type and frequency of inspection to be used depend on the control of the manufacturing process (reliability).
- There are often different but equivalently correct inspection methods. Prescription of particular inspection methods would force the manufacturer to provide inspection devices prescribed by the customer, although other sufficient inspection devices are already available.

• Prescribing inspection methods that differ in assessment from the precise tolerance zone requires further specifications of the measuring conditions. Inspection methods that differ from the precise tolerance zone and different measuring conditions would make the inspection of geometrical deviations obscure and prone to mistakes.

Therefore the ISO Standards do not recommend the indication of a particular inspection method. The drawing shall only specify the geometrically exactly defined requirement (tolerance zone) (see 19).

However, ISO intends to prepare standards on the measurement of form deviations. The standards will specify the ideal operators (nominal properties of the reference measuring process) that are as close as practically possible to the geometrical exact definition of the deviation. The planned standards are:

ISO 12 780	Measurement of straightness deviations
ISO 12 781	Measurement of flatness deviations
ISO 12 181	Measurement of roundness deviations
ISO 12 180	Measurement of cylindricity deviations

These methods are then the bases to which the measurement uncertainties refer. A measurement with an ideal operator would have the measurement uncertainty zero. (But this measurement is practically impossible because each realization of a measuring instrument has deviations from the ideal operator (properties) and produces therefore a measurement uncertainty.)

18.2 Terms

Embodiment: Measuring is comparing. In order to measure geometrical deviations, the workpiece surface must be compared with a geometrical ideal feature. But geometrical ideal features cannot be manufactured. Therefore almost geometrical ideal embodiments of geometrical ideal features (e.g. straight lines, planes, circles, cylinders and spheres) are used. They are called embodiments in the following.

The embodiments can be surfaces of measuring devices (e.g. straight edges, measuring tables, solid angles, sliding guides of measuring devices) or can be established by the movements of precision guides (e.g. by rotating the workpiece or the measuring device).

Datum: Theoretically exact geometrical reference (such as an axis, plane or straight line) to which toleranced features are related. Datums may be based on one or more datum features of a workpiece; ISO 5459 (see 3.4).

Datum system: Group of two or more separate datums used as a combined reference for a toleranced feature; ISO 5459 (see 3.4).

Datum feature: Real feature of a workpiece (such as an edge, a plane surface or a hole) that is used to establish the location of a datum; ISO 5459 (see 3.4).

Datum target: Point, line or limited area on the workpiece to be used for contact with the manufacturing and inspection equipment, to define the required datums in order to satisfy the functional requirements; ISO 5459 (see 3.4).

Simulated datum feature: Real surface of sufficiently precise form (such as a surface plate, a bearing, or a mandrel) contacting the datum feature(s) and used to establish the datum(s).

Simulated datum features are used as the **practical embodiment** of the datums during manufacture and inspection; ISO 5459 (see 3.4).

Reference element: Geometrical ideal element (straight line, circle, line of any form defined by theoretical exact dimensions, plane, cylinder, or a surface of any form defined by theoretical exact dimensions) relative to which the geometrical deviations are evaluated.

The reference element is established by or derived from the embodiment at the measured (toleranced) feature. When the reference element is calculated (e.g. using a coordinate measuring machine), the reference element may differ from the embodiment in orientation, location and size.

The orientation and location of the reference element depend on the characteristic to be measured (deviations of unrelated profile (form), related profile, orientation, circular or total axial run-out, location, circular or total radial run-out); see below, Geometrical deviation.

Coordinate measuring machines and form measuring instruments often approximate the reference element by the (Gaussian) least squares substitute element (see 8.1).

The substitute element intersects or contacts the workpiece surface, but the reference element need not (see Fig. 8.5 and Table 18.1). In cases of a straight line or plane the minimum zone substitute element and the reference element have the same orientation but not necessarily the same location. In cases of a circle or cylinder the minimum zone substitute element and the reference element have the same location and orientation but not necessarily the same size.

Geometrical deviation: The definitions of geometrical deviations are not yet internationally standardized (International Standards on this subject are planned). But they can be derived from the definitions of geometrical tolerances according to ISO 1101.

The geometrical deviation is the deviation of a workpiece feature (axis, section line, edge, surface or median face) from the (geometrical ideal) reference element (embodiment).

The reference element is

- orientated according to the minimum requirement (18.3.2) for the measurement of deviations of unrelated profile (form);
- orientated according to the datum (18.3.4) or datum system (3.4) for the measurement of deviations of related profile, orientation, circular or total axial run-out;
- orientated and located according to the datum (18.3.4) or datum system (3.4) and to the theoretical exact dimensions for the measurement of deviations of related profile, location, circular or total radial run-out.*

For the evaluation of the geometrical deviation to be compared with the geometrical tolerance, see 18.6.

18.3 Alignment of the workpiece

18.3.1 General

Most important for measurement is the alignment of the workpiece in the measuring device. Misalignment can cause large errors in measurement, so that workpieces that

---

* In the cases of circular radial run-out or total radial run-out the theoretical exact dimension between the axes of the toleranced feature and the datum feature is zero. For cases of related profile see Fig. 4.4.

actually comply with the specification may appear not to, more seldom but in principle possibly, workpieces that actually do not comply with the specification may appear as complying.

For the measurement of form deviations the minimum requirement has to be respected (see 18.3.2 and 18.3.3).

For the measurement of deviations of orientation, location and run-out (related geometrical deviations) the minimum rock requirement has to be respected for the datum(s) (see 18.3.4).

For the measurement of roundness or cylindricity additional alignment requirements have to be respected (see 18.3.3).

18.3.2 Minimum requirement

The minimum requirement defines the orientation or location of the form tolerance zone.

With straightness and with flatness, it requires that the parallel straight lines or parallel planes enclosing the feature be directed so that their distance is a minimum (Fig. 18.1); ISO 1101.

With roundness and cylindricity, it requires that the concentric circles and the coaxial cylinders enclosing the feature be located so that their radial distance is a minimum (Fig. 18.2); ISO 1101.

Deviations from the minimum requirement (e.g. alignment according to the Gaussian regression line) simulate larger form deviations.

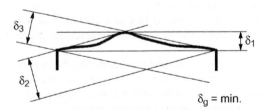

Fig. 18.1 Minimum requirement for straight lines

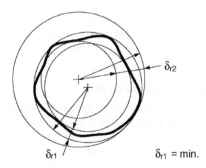

Fig. 18.2 Minimum requirement for circles

18.3.3 Additional alignment requirements for the measurement of roundness or cylindricity deviations

18.3.3.1 Axis of measurement

The **axis of measurement (reference axis)** is a straight line relative to which the measurement is performed and established by the measuring device. The measurements are considered to be perpendicular to and centred on the axis of measurement.

The (desired) **workpiece established axis** is a straight line about which the relevant part of the workpiece is required to surround and may be defined (established) as follows.

(a) A straight line such that the root mean square value of the distances from it to the defined centres of a representative number of cross-sections has a minimum value. The defined centres may be the centres of the
  - least squares circle, LSC (according to Gauss, the circle of which the sum of the squares of the radial distances to the circumference is a minimum, regression circle);
  - minimum zone circle, MZC (according to Chebyshev, the circle of which the maximum radial distance to the circumference is a minimum);
  - contacting circle (maximum inscribed circle for holes, MIC; minimum circumscribed circle for shafts, MCC; sometimes referred to as plug gauge circle, PGC, and ring gauge circle, RGC).

(b) A straight line passing through the defined centres of two separated and defined cross-sections. For the defined centres see (a).

(c) A straight line passing through the defined centre of one defined cross-section and perpendicular to a defined shoulder. For the defined centres see (a). The defined shoulder may be determined by the
  - least squares plane (according to Gauss, the plane of which the sum of the squares of the distances to the actual feature (shoulder) is a minimum, regression plane);
  - minimum zone plane (according to Chebyshev, the plane of which the maximum distance to the actual feature (shoulder) is a minimum);
  - datum plane (according to the minimum rock requirement, see 18.3.4).

(d) A straight line passing through two support centres.

(e) An axis of two coaxial cylinders with minimum separation to the surface irregularities of the workpiece (minimum zone).

(f) A substitute cylinder axis (see 8.1).

If not otherwise specified (agreed upon) according to ISO 1101, the definition (e) applies. However, current practice is, for practical reasons, to use one of the other possibilities, preferably one of the Gaussian methods.

18.3.3.2 Inclination of the workpiece axis

If the (desired) workpiece established axis is inclined to the axis of measurement (reference axis) a round workpiece feature (about the desired workpiece established axis) may appear oval and an oval workpiece feature may appear round (Fig. 18.3).

The planned ISO Standards on measurement of roundness and cylindricity will probably specify that the levelling (alignment) should be performed at the top and bottom

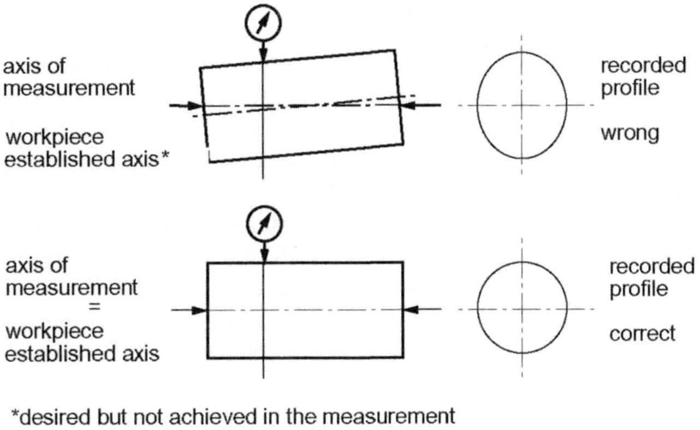

axis of
measurement

workpiece
established axis*

recorded
profile

wrong

axis of
measurement
=
workpiece
established axis

recorded
profile

correct

*desired but not achieved in the measurement

Fig. 18.3 Roundness measurement: false measurement caused by a plane of measurement not perpendicular to the workpiece established axis (workpiece established axis inclined to the axis of measurement)

of the workpiece feature until the least squares centres (LSC) at these two levels coincide to a value better than 10% of the roundness or cylindricity tolerance.

18.3.3.3 Offset of the workpiece axis

Some form measuring instruments highly magnify the form deviations but not the radius itself. When these instruments are used and when in the measuring plane the centre of the workpiece cross-section line (profile) differs from the axis of measurement (reference axis), distortions of the profile occur (limaçon effect) (Fig. 18.4).

The planned ISO Standard on the measurement of roundness will probably specify that the workpiece should be centred at each level that is measured individually to improve the resolution of the measurement and minimize the limaçon error. The centring should be continued until a measuring range can be used (i.e. the profile is fully contained in this range) that results in a resolution of the instrument that is smaller than 1% of the roundness tolerance.

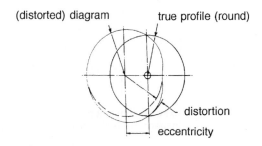

(distorted) diagram   true profile (round)

distortion

eccentricity

Fig. 18.4 Distortion of the profile diagram caused by an offset of the true workpiece centre from the axis of measurement (limaçon effect)

18.3.4 Minimum rock requirement

In the measurement of deviations of orientation, location and run-out, the minimum rock requirement applies to the alignment of the datum feature.

When the datum feature is not stable (convex datum feature) relative to the contacting surface (simulated datum feature, e.g. measuring table or mandrel) it shall be arranged so that the possible movement (inclination) in any direction is equalized, i.e. that the maximum possible inclination to the extreme position is a minimum (minimum rock requirement) (Fig. 18.5 and 18.6). In other words, the datum feature shall be aligned relative to the simulated datum feature into a median position.

In the case of a planar datum feature an approximation to the minimum rock requirement is to place three equal-height supports between the datum feature and the simulated datum feature at the same distances from the end of the datum feature.

In the case of a cylindrical datum feature (axis) the contacting cylinder (the maximum inscribed simulated datum cylinder for a datum feature that is a hole, and minimum circumscribed simulated datum cylinder for a datum feature that is a cylindrical shaft) shall be arranged according to the minimum rock requirement, i.e. in a median position (Fig. 18.6).

When the datum feature is the common axis of two cylindrical shafts of different nominal sizes according to ISO 5459, the datum is the common axis established by the two smallest circumscribed coaxial cylinders (Fig. 18.7). This definition is not precise. Figure 18.7 is taken from ISO 5459, and shows the case in which the **sum** of the diameters of the coaxial simulated datum cylinders is a minimum.

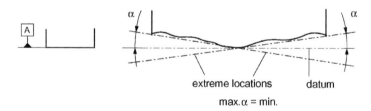

extreme locations   datum

max. α = min.

Fig. 18.5 Minimum rock requirement for planar datum features

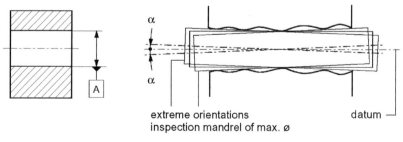

extreme orientations   datum
inspection mandrel of max. ø

max. α = min.

Fig. 18.6 Minimum rock requirement for cylindrical datum features

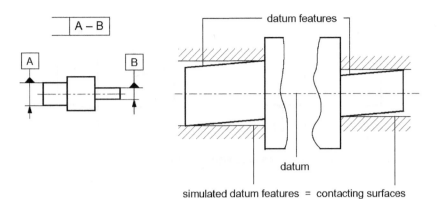

simulated datum features = contacting surfaces

Fig. 18.7 Common axis as a datum (ISO 5459)

A possible definition of a common axis of two or more datum features (which is under consideration for the revision of ISO 5459) is the axis of the smallest imaginary cylinder containing the actual axes of the datum features. Another possibility under consideration is the definition of two coaxial cylinders with maximum distance to the workpiece minimized (minimum zone).

Often the connection of the centres (centres of the least squares circles) of the cross-sections at half the length of the datum features is used to establish the common axis (Fig. 18.8) (measurement with form measuring instrument or support in edge V-blocks). See also 18.7.10.1.

In order to avoid disputes, when small geometrical tolerances and common axes are required, the method for establishing the common axis should be specified in the drawing, e.g. support at specified cross-sections (Fig. 18.8).

For the alignment of the datum features according to the three-plane concept, see 3.4.

The minimum rock requirement may cause difficulties when the actual surface of a planar datum surface is convex and, for example, more inclined near one of the edges. Then the extreme rock positions and the minimum rock position may be unreasonable. Discussions are presently under way in the ISO whether the minimum rock requirement should be replaced by, for example, the minimum zone requirement. ASME Y14.5 defines the candidate datum method. This method selects contacting planes which contact the actual datum feature to a certain extent and give possible orientations relative to which all related tolerances are to be respected.

18.4   Interchanging the datum feature and toleranced feature

With related geometrical tolerances, the drawing distinguishes between toleranced features and datum features. When, for the inspection, toleranced and datum features are interchanged, completely different values of the geometrical deviation may be obtained from the same workpiece (Fig. 18.9).

Therefore, for the inspection, toleranced and datum features must not be interchanged. When the datum feature indicated on the drawing is not suitable for inspection purposes (e.g. because it is too short) a change in the drawing is necessary (Figs 19.13 and 19.14).

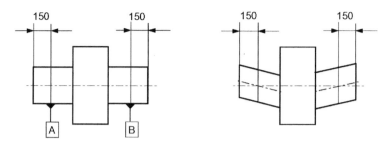

Fig. 18.8 Common datum axis established by the centres of two specified sections

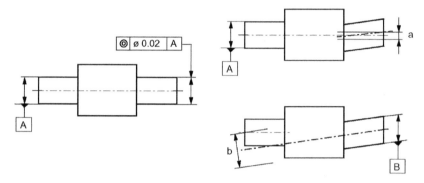

Fig. 18.9 Change in geometrical deviation caused by interchanging of toleranced and datum features

18.5 Simplified inspection method

Inspection of certain types of geometrical tolerances (e.g. coaxiality tolerance) is relatively costly. Often in such cases in the first step a "quick" (but less precise) inspection is chosen and only in case of doubt is the more precise (and more costly) inspection executed in a second step.

For example, in the cases of a coaxiality tolerance or a common straightness tolerance zone of axes, the inspection in the first step is performed as if there were a run-out tolerance of the same value. Only when this tolerance is exceeded is it checked by a more precise method whether the coaxiality tolerance or the straightness tolerance of the axis is exceeded.

Similar considerations apply in the cases of a cylindricity tolerance and a roundness tolerance. In the first step the check of run-out may be executed.

Often with related geometrical tolerances, the precise verification of the minimum rock requirement is very costly. Therefore approximate inspection methods are used with the aid of V-blocks, mandrels, centre holes, etc. Using V-blocks, the form deviation of the datum feature leads to simulation of a larger related geometrical deviation than actually exists (according to the definition). Depending on the shape of the form deviation and the angle of the V-block, the increase can be equal or less than the form deviation of the datum feature (see 18.7.4.3).

A similar consideration applies with the use of centre holes. There the eccentricity of the centre hole relative to the datum feature increases the result of the measurement of the related geometrical deviation.

Mandrels for inspection purposes are normally rated in diameter in units of 0.01 mm. The largest mandrel that fits into the hole is to be used. In cylindrical holes the mandrel can be inclined by 0.01 mm at most (Fig. 18.10), and thereby give an incorrect measuring result. In very rare cases of very detrimental form deviations larger inclinations may occur (Fig. 18.11).

When the indicated excess of the tolerance is of the same magnitude as the possible error caused by the inspection equipment and significant compared with the tolerance,

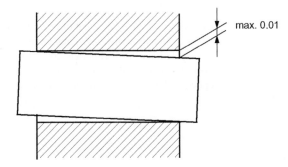

Fig. 18.10 Mandrel: possible inclination

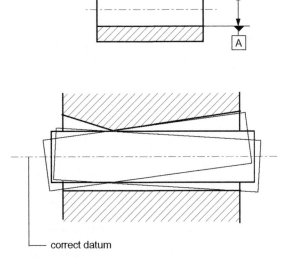

Fig. 18.11 Mandrel: possible deviation from the minimum rock requirement

more precise inspections should be executed (or the inspection method should be agreed upon between the parties). See also 18.11.12.

Pneumatic mandrels are self-centring, as are expanding mandrels and conical mandrels.

18.6 Evaluation of measurement

Evaluation of a measurement can be done manually by calculation, or graphically or automatically by suitable computer programs (e.g. in a form measuring instrument or in a coordinate measuring machine). In the following sections the principles for the evaluation by calculation are described.

In many cases a cylindrical tolerance zone is derived from the function, and is indicated in the drawing. When during inspection the deviations in Cartesian (rectangular) coordinates are assessed, they can be compared with the tolerance diameter with the aid of the diagram in Fig. 18.12.

Figure 18.13 shows a typical method of the assessment of related geometrical deviations (here parallelism deviations of an axis relative to a datum axis) by detecting rectangular coordinates δ_x, δ_y and calculation of the cylindrical coordinate δ_p.

Often the tolerance applies over the length l of the feature, whereas for practical reasons the measurement was applied to the length l_m. The measured values are to be corrected by the ratio l/l_m, i.e.

$$\delta_u = \delta_m \, (l/l_m)$$

where
δ_u = geometrical deviation
δ_m = measured deviation over the length l_m

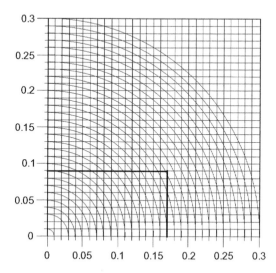

Example: measured coordinates 0.17 and 0.09 lie within ø 0.4 ($r = 0.2$)

Fig. 18.12 Diagram to compare assessed coordinates with cylindrical or circular tolerances

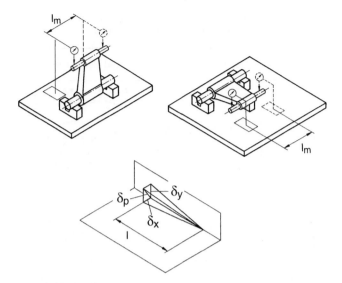

Fig. 18.13 Measuring of the radial deviation of parallelism by detecting the deviations in rectangular coordinates

l = length of the feature
l_m = distance between the measured points

The geometrical tolerances t are defined as widths of tolerance zones within which the toleranced feature must be contained. Geometrical deviations δ, however, are measurable deviations from the geometrical ideal form δ_f, orientation δ_d or location δ_o.

The maximum permissible deviation from location δ_o (positional deviation δ_o, coaxiality deviation δ_a, and symmetry deviation δ_s) corresponds to one-half of the value of the location tolerances, $\delta_o = t_o/2$ (Figs 18.52 and 18.53, compare Figs 18.56 and 18.57 with Fig. 3.4). The same applies to deviations of the profile of a line, δ_b, or of a surface, δ_h, defined by theoretical exact (rectangular framed) dimensions (Figs 4.1 to 4.3). See Table 18.1.

In all remaining cases the maximum permissible geometrical deviation corresponds to the full value of the geometrical tolerance (Figs 18.48 to 18.51). See Table 18.1.

18.7 Methods of inspection

The following provides a survey of the most relevant inspection methods.

The former East German Standards TGL 39 093 to TGL 39 098 and TGL 43 041 to TGL 43 045 and TGL 43 529, TGL 43 530 provide more detailed descriptions of the same inspection methods. However, it should be noted that tolerances of orientation and location according to TGL 19 080 and according to the former Comecon Standard (in east and middle Europe) ST RGW 301-76 are defined by eliminating the form deviations of the toleranced feature, i.e. deviations of orientation and location do not include deviations of form (Fig. 18.14). This is in contrast to ISO 1101.

Table 18.1 Synopsis of geometrical deviations δ and comparison with geometrical tolerances _t_

characteristic		reference: — - — - — toleranced element: ———
form	⌒ – ○ ⌓ ⌗ ⌰	$t \geq \delta = A_{max} - A_{min}$
orientation	∠ // ⊥	$t \geq \delta = A_{max} - A_{min}$
location	⊕ ◎ ⊜ ⌒ ⌓	$t \geq 2\delta = 2A_{max}$
run-out	↗ ↗↗	$t \geq \delta = A_{max} - A_{min}$

In the following the principles of the inspection methods corresponding to ISO 1101 are described.

18.7.1 Assessment of straightness deviations of lines of surfaces

18.7.1.1 Definition

The deviations of the line of the workpiece surface from an (almost) geometrical ideal reference straight line (embodiment, e.g. established by a straight edge) are measured. The line of the workpiece surface and the reference line are contained in a section plane (Fig. 3.15), or in a projection plane (i.e. a plane onto which the lines and the two parallel planes establishing the tolerance zone are projected) (Figs 3.19 and 3.22).

When the reference line does not intersect (but eventually touches) the workpiece line, the straightness deviation δ_g according to ISO 1101 is the difference between the largest and smallest distances between the workpiece line and the reference line (Fig. 18.16).

When the reference line does intersect the workpiece line, the straightness deviation according to ISO 1101 is the sum of the largest distances of the workpiece line from

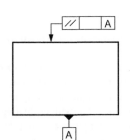

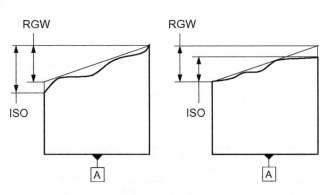

Fig. 18.14 Deviation (tolerance) of orientation and location according to ISO 1101 and according to ST RGW 301-76 and ST RGW 368-76

the reference line on both sides of the reference line (Table 18.1), i.e. the range of the local deviations of the workpiece line from the reference line (Fig. 8.28).

The reference line is to be aligned according to the minimum requirement (Fig. 18.1).

The straightness deviation δ_g must not exceed the straightness tolerance t_g: $\delta_g \leq t_g$.

A standard (ISO 12 780) on the measurement of straightness deviations is in preparation.

18.7.1.2 Type of detecting

The deviations can be detected by

1. continuously probing and recording;
2. consecutive probing (sampling, approximation) and recording

while

(a) measuring the distance to the reference line (embodiment);
(b) measuring the inclination of a two point bridge on the surface relative to the reference line (embodiment).

18.7.1.3 Measuring methods

The reference line (embodiment) may be established by:

- straight guides   (of a form measuring instrument)
  (of a coordinate measuring instrument);
- measuring plate   (with length measuring instrument, e.g. dial gauge)
  (optical flat with evaluating of contour lines);
- straight edge   (with length measuring device, e.g. dial gauge or feeler gauge);
- measuring wire   (with gauge);
- straight line marking   (within profile projector);
- optical axis   (collimating microtelescope with targets)*
  (autocollimator and test mirror)†
  (laser beam with photoelectric detector)*
  (laser interferometer with two-point bridge)†
- Earth curvature   (liquid surface of a hose levelling instrument)*‡
  (two-point bridge with inclinometer)†‡

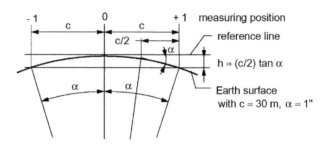

Fig. 18.15 Height correction h in a distance c for eliminating the effect of the Earth's curvature

When the surface (workpiece line) is concave, the reference line (embodiment) aligns automatically according to the minimum requirement (Fig. 18.16).

When the surface (workpiece line) is more complicated or longer an approximation is that both ends have equal distance from the reference line (Fig. 18.17).

When form measuring instruments or coordinate measuring machines are used, the alignment, as an approximation, is often performed according to the Gaussian regression line (sum of squares of the distances to the regression line is a minimum). Then the obtained value of the straightness deviation can be larger than the value according to ISO 1101 (minimum zone value).

---

* Height measurement.
† Inclination measurement.
‡ With long reference lines the curvature of the Earth ($\approx 1''$ per 30 m) must eventually be taken into account (Fig. 18.15, Table 18.2).

Table 18.2 Height correction _h_ over a distance _c_ for eliminating the effect of the Earth's curvature (see Fig. 18.15)

_c_ (m)	_h_ (μm)	_c_ (m)	_h_ (μm)	_c_ (m)	_h_ (μm)
1	0.08	11	9.8	25	50
2	0.32	12	11	30	73
3	0.73	13	14	35	99
4	1.3	14	16	40	129
5	2.0	15	18	45	160
6	2.9	16	21	50	203
7	3.9	17	23	60	291
8	5.2	18	26	70	396
9	6.5	19	29	80	517
10	8.0	20	32	100	800

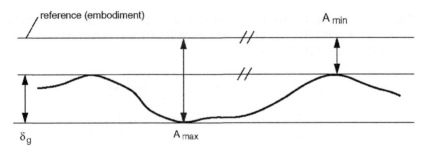

Fig. 18.16 Measurement of straightness deviation; alignment according to the minimum requirement

When the toleranced length is longer than the straightness embodiment, the embodiment (e.g. straight edge with height adjustable supports) is used in consecutive measuring positions without (a) or with (b) overlapping.

(a) When the straightness embodiment is used in positions without overlapping, it is to be adjusted according to a levelling instrument. The first height indication of the following position is to be adjusted according to the last height indication of the preceding embodiment position (Fig. 18.18).

The adjustment of the connecting points of the straight edge during measurement can be omitted and be transferred into the graphical evaluation (Fig. 18.19).

(b) When the straightness embodiment is used in overlapping positions, at least two measuring positions must overlap. The two overlapping height indications of the

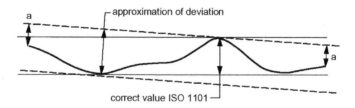

Fig. 18.17 Measurement of straightness deviation; continuous lines exact alignment according to the minimum requirement: dashed lines, approximation by equal distance *a* at both ends

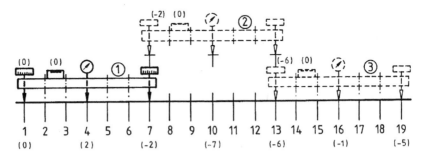

Fig. 18.18 Consecutive measurement of straightness deviation with adjusted straight edge without overlapping, example with measured values in parentheses, according to TGL 39 093

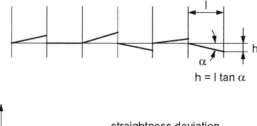

$$h = l \tan \alpha$$

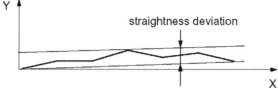

Fig. 18.19 Graphical evaluation of the straightness deviation from consecutive measurement of the deviation from the horizontal orientation

following position have to be adjusted according to the height indications of the preceding embodiment position (alignment of the straight edge) (Figs 18.20 and 18.21). The measurement uncertainty is lower the more the measuring positions overlap.

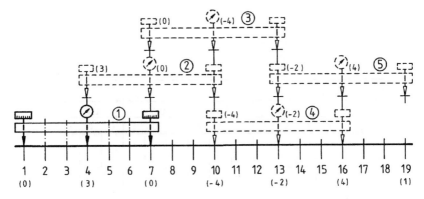

Fig. 18.20 Measurement of straightness deviation with overlapping straight edges, example with measured values in parentheses, according to TGL 39 093

When the angle α (inclination) of the connection of adjacent measuring positions relative to the reference line (straightness embodiment) is measured in seconds, the height difference is $h = c \tan \alpha$ in µm, where c is the distance in m between the measuring positions ($\tan 1'' = 4.848$ µm/1000 mm ≈ 5 µm/1000 mm) (Fig. 18.22).

18.7.2 Assessment of straightness deviations of axes

18.7.2.1 Definition

With cylindrical features, the actual axis can be taken as a sequence of centres of circles that can be defined as follows:

(a) least squares circle, LSC (according to Gauss, the circle of which the sum of the squares of the radial distances to the circumference is a minimum, regression circle);
(b) minimum zone circle, MZC (according to Chebyshev, circle of which the maximum radial distance to the circumference is a minimum);
(c) contacting circle (maximum inscribed circle for holes, MIC, minimum circumscribed circle for shafts, MCC, sometimes referred to as plug gauge circle, PGC, and ring gauge circle, RGC);

in cross-sections perpendicular to the axis of the following reference cylinders:

(a) least squares cylinder (according to Gauss, the cylinder of which the sum of the squares of the radial distances to the actual cylinder surface is a minimum);
(b) median cylinder of minimum zone cylinder (according to Chebyshev, cylinder of which the maximum radial distance to the actual cylinder surface is a minimum);
(c) contacting cylinder (maximum inscribed cylinder for holes and minimum circumscribed cylinder for shafts);
(d) two cross-section cylinder, defined by the least squares circles or the minimum zone circles or the contacting circles of two cross-sections near the ends of the feature.

ISO 14 660-2 defines as a default the Gauss solutions (a) and (a), see also 8.3.1.

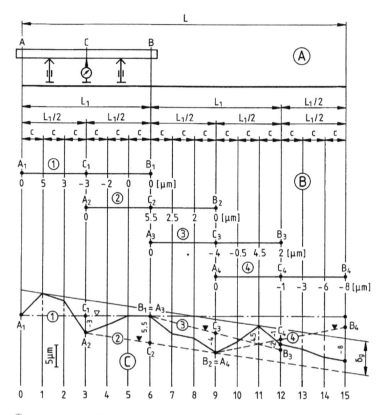

(A) measuring scheme and length of measuring steps
(B) positions of straight edge and examples of distances to the straight edge
related to position A of the straight edge
(C) examples of graphical evaluation of the measuring result

Fig. 18.21 Graphical evaluation of the straightness deviation for consecutive measurement of the deviations using overlapping positions of the straight edge, example with measured values, according to TGL 39 093

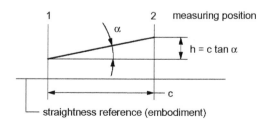

Fig. 18.22 Inclination α along a distance c converted to height deviation h

Coordinate measuring machines apply normally the least squares cylinder and the least squares circle and, if optionally available, the minimum zone circle and the contacting circles. Form measuring instruments apply normally the two cross-sections cylinder defined by the least squares circles or minimum zone circles or contacting circles. (But so far with coordinate measuring machines the reference cylinder is predominantly the least squares cylinder, because the mathematical techniques have been developed early, while with form measuring instruments it is normally the two cross-sections cylinder.)

The least squares circle and the least squares cylinder are the only methods that are always unique and that need the least number of measurements (probes, measured points) to be sufficiently stable in size and in location. Therefore they have become the default case (to be applied if not otherwise specified) according to ISO 14 660-2.

The deviations of the actual axis from an (almost) geometrical ideal reference straight line (embodiment) are measured. The deviations have to be calculated in relation to an (imagined) reference line to which the minimum requirement applies, i.e. the maximum local deviation e must be a minimum (Figs 18.1 and 18.24). The straightness deviation δ_g is the range of the local deviations e: $\delta_g = 2e_{max}$ (Figs 8.28 and 18.24). The straightness deviation must not exceed the straightness tolerance t_g, i.e. the maximum local straightness deviation e_{max} must not exceed one-half of the straightness tolerance t_g:

$$\delta_g = 2e_{max} \leqslant t_g$$

A standard (ISO 12 780) on the measurement of straightness deviations is in preparation.

18.7.2.2 Assessment of the coordinates of axes

Depending on the drawing indication, the straightness tolerance is defined as the zone in a projected plane in order to assess the curvature projected in this plane (Fig. 3.22(a)) or as the zone (distance) between two parallel planes (Fig. 3.22(b)) or as a cylindrical zone in order to assess the curvature in space (Fig. 3.17), see 3.3.1.

With tolerance zones in projected planes or between parallel planes, the evaluation of the deviation is similar to that described in 18.7.1.

With a cylindrical tolerance zone, i.e. for the assessment of the straightness deviation (curvature) in the space, it is normally measured in each cross-section in two mutually perpendicular directions (coordinates x_i, y_i, z_i, Fig. 18.23).

The smallest cylinder containing all assessed centres, is to be determined.

There are approximations for this purpose.

One approximation is to define a reference line (z direction) to align the workpiece in this coordinate system, to assess the actual axis (coordinates) and to calculate the distances between the coordinates of the actual axis and the reference line. The local distance in space is $e_i = \sqrt{(x_i^2 + y_i^2)}$, where

x_i = the distance in the xz plane (x direction)
y_i = the distance in the yz plane (y direction)

The diameter of the smallest cylinder containing all centres (points of the actual axis) is $\delta g = 2e_{max} \leqslant t_g$. Its axis is the reference line (Fig. 18.23).

Often the reference line is established by the centres of the first and last cross-sections. With coordinate measuring machines, the reference line is the axis of the substitute cylinder (least squares cylinder) (see 8, 18.7.2.1 and 18.3.3).

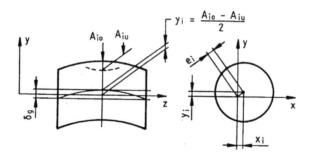

Fig. 18.23 Assessment of coordinates of an actual axis (centres of cross-sections)

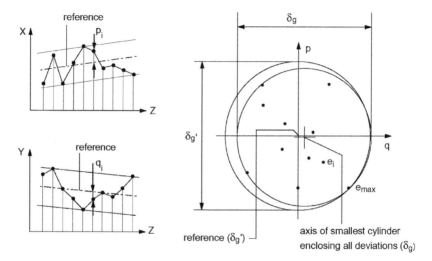

Fig. 18.24 Evaluation of the straightness deviation of an axis in space

When the reference line is not exactly parallel to the z axis, but only approximately so, the measured coordinates x_i can be recorded in an xz coordinate system and y_i in a yz coordinate system (Fig. 18.24). For an approximation in each coordinate system the reference line is to be determined so that the maximum deviation is a minimum (parallel to the two parallel straight lines of minimum distance that enclose all points of the actual axis; see 18.7.1.3). The deviations p and q of the points from the reference line are to be determined. The local actual deviations in space from the reference line are $e_i = \sqrt{(p_i^2 + q_i^2)}$, and the diameter of the smallest cylinder that contains all centres e_i (points of the actual axis) is the straightness deviation δ_g. The straightness deviation δ_g must not exceed the straightness tolerance t_g: $\delta_g \leq t_g$.

An approximation, easy to calculate, is (Fig. 18.24)

$$\delta_g \approx \delta_g' = 2e_{max} \leq t_g.$$

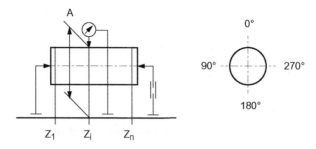

Fig. 18.25 Assessment of the straightness deviation of an axis with one dial indicator

18.7.2.3 Assessment with one dial indicator

A simple approximate method for assessing the straightness deviation of an axis is shown in Fig. 18.25. The workpiece has to be supported centrally (e.g. in the centres of the measuring positions z_1 and z_n).

In each cross-section (measuring position) the coordinates of the centres are to be determined by probing at the angle positions 0° and 180°, 90° and 270°. The coordinates of the centre (observing the signs + and −) are to be introduced into a polar diagram (in the x direction $x_i = (A_0 - A_{180})/2$ and in the y direction $y_i = (A_{90} - A_{270})/2$) (Fig. 18.23). The diameter of the smallest circle enclosing the centres of all cross-sections corresponds to the straightness deviation δ_g.

With this method, problems arise with positioning the axis of revolution (reference line) according to the minimum requirement and with the assessment of the actual centres according to a definition given in 18.7.2.1.

18.7.2.4 Assessment with two dial indicators

Another simple approximation method for assessing of the straightness deviation of an axis is shown in Fig. 18.26. The workpiece axis has to be aligned parallel to the reference line (embodiment, e.g. measuring table), for example through the centres of the measurement positions z_1 and z_n with the same distance from the measuring table.

In each longitudinal section (containing the axis), of e.g. 4 sections, the values $R = (A_0 - A_u)/2$ are to be determined at several (at least three) measuring positions. The difference between R_{max} and R_{min} within one section represents the straightness deviation of the axis in this section. The straightness deviation of the axis of the cylindrical feature is the maximum of the straightness deviations of the sections (Fig. 18.27):

$$\text{Straightness deviation of the axis } \delta_g = \left(\frac{A_0 - A_u}{2}\right)_{max} - \left(\frac{A_0 - A_u}{2}\right)_{min} \leq t_g$$

As the directions of measurement are opposite, it is the difference of half the sums of the distances from the measuring plate that is determined.

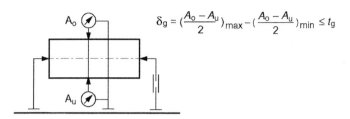

$$\delta_g = \left(\frac{A_o - A_u}{2}\right)_{max} - \left(\frac{A_o - A_u}{2}\right)_{min} \leq t_g$$

Fig. 18.26 Assessment of the straightness deviation of an axis with two dial indicators

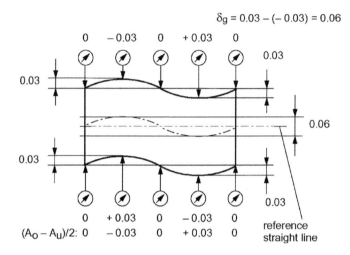

Fig. 18.27 Example of the assessment of the straightness deviation of an axis with two dial indicators

For the measurement of the straightness deviations of generatrixes (generator lines) or axes of cylindrical or conical shafts (or edges) cylindrical or flat anvils should be used (see 3.3.1 and Figs 3.19 and 18.117(c)).

18.7.2.5 Assessment with form measuring instrument or with coordinate measuring machine

The instruments assess the coordinates of points of the actual circumference in cross-sections approximately perpendicular to the axis and calculate the coordinates of the actual centre points of which the actual axis is composed. From the points of the actual axis, they calculate the local straightness deviations, taking into account the minimum requirement. Twice the value of the maximum local straightness deviation corresponds to the straightness deviation δ_g according to ISO 1101 (see 18.7.2.2).

The instruments allow measurements close to the definitions. For the definition of the actual axis, see 18.7.2.1.

18.7.3 Assessment of flatness deviations

18.7.3.1 Definition

The deviations of the surface from an (almost) ideal reference plane (embodiment, e.g. established by a measuring plate) are measured.

When the reference plane does not intersect (but eventually touches) the workpiece surface, the flatness deviation δ_e according to ISO 1101 is the difference between the largest and smallest distances between the workpiece surface and the reference plane (Table 18.1).

When the reference plane does intersect the workpiece surface, the flatness deviation δ_e according to ISO 1101 is the sum of the largest distances between the reference plane and the workpiece surface above and below the reference plane (Table 18.1).

The reference plane is to be aligned according to the minimum requirement (Fig. 18.1). The flatness deviation δ_e must not exceed the flatness tolerance t_e: $\delta_e \leq t_e$.

A standard ISO 12 781 on the measurement of flatness deviations is in preparation.

18.7.3.2 Type of detection and measurement methods

These are similar to those used for the assessment of straightness deviations (see 18.7.1.2 and 18.7.1.3).

When the deviations are assessed in lines with the aid of straightness embodiments (e.g. with a straight edge), the embodiment must remain in the same plane (inclination, height level) or the measured values must accordingly be corrected by calculation.

Figure 18.28 shows a measuring device where the alignment is controlled by an inclination measuring instrument.

However, it is difficult to align the workpiece surface relative to the plane embodiment so that the largest measured deviation is a minimum. Often, as an approximation, the distances between workpiece surface and plane embodiment are equalized at three ends (points) of the surface (e.g. three supports of equal height). Another approximation, when a computer is used (e.g. with coordinate measuring machines), is alignment parallel to the least squares plane (the plane of which the sum of the squares of the distances to the actual surface is a minimum). With the approximations, there are always larger estimations of the deviations than according to the minimum requirement according to ISO 1101.

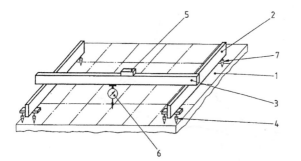

1 workpiece
2 straight edge
3 straight edge
4 adjustable supports
5 inclinometer
6 dial gauge
7 adjustable supports

Fig. 18.28 Assessment of flatness deviation of a workpiece (1) with straight edges (2, 3), adjustable supports (4, 7), inclinometer (5) and dial gauge (6)

The plane embodiment according to the minimum requirement has one of the following positions (alignments):

(a) it touches the three highest points (concave form);
(b) it touches the three deepest points (convex form);
(c) it touches the two highest points and is parallel to the straight line touching the two deepest points (saddle form, Fig. 18.29).

The former East German standard TGL 39 094 describes a graphical method "ZNI-ITM" for the assessment of the flatness deviation close to the definition (minimum requirement) as follows:

The distances of the workpiece surface from the plane embodiment are measured. The inclination of the embodiment relative to the workpiece surface should be small.

The measurement positions (points) are enumerated and plotted on a scale, Ⓐ in Fig. 18.30, measuring points A1 to C5. At each point the measured value is indicated. In the view Ⓑ are plotted on the scale the height differences from the plane embodiment. The straight line P_x touches the points from above and is aligned with respect to the points according to the minimum requirement. P_x determines one direction of the reference plane.

In an inclined plane Ⓒ, perpendicular to the straight line P_x (connection B1–A4), the distances of the surface points from P_x (in view Ⓑ) are plotted (e.g. a). The straight line P_y touches the highest point and is directed according to the minimum requirement (δ_e = min). P_y determines the other direction of the reference plane. The point with the largest distance (in the direction of P_x, B1–A4) to P_y corresponds to the flatness deviation δ_e (point A1). (The reference plane in Fig. 18.30 is determined by the points B1, C3 and A4).

Another method according to TGL 39 094 for the assessment of the flatness deviation close to the definition uses analogue mechanical devices (Fig. 18.31).

The pins are movable and to be aligned on a larger scale according to the measured deviations. As all pins are of equal length, the other side exhibits a mirror image of the measured form. By inclining the device on a measuring plate, the position is to be determined in which the pin with the largest distance to the measuring plate has the least distance. This distance corresponds to the flatness deviation δ_e.

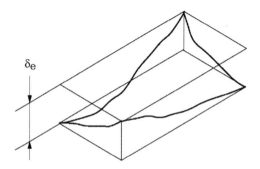

Fig. 18.29 Flatness deviation δ_e of a saddle form surface

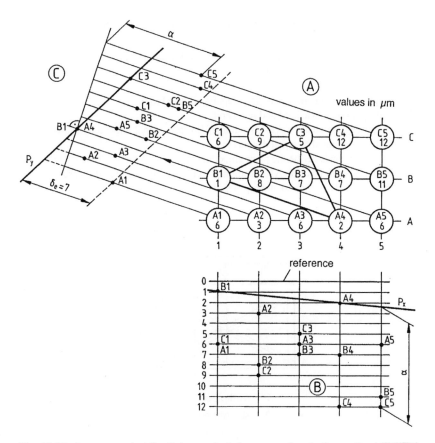

Fig. 18.30 Assessment of the flatness deviation according to the method ZNIITM

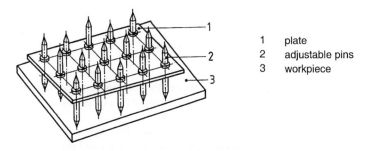

1	plate
2	adjustable pins
3	workpiece

Fig. 18.31 Analogue mechanical device for the assessment of the flatness deviation

18.7.3.3 Assessment of the flatness deviation with straight edge and dial indicator

The procedure of the assessment of the flatness deviation with straight edge and dial indicator (Fig. 18.32) is as follows:

1. straight edge in position C1–A5, supports adjusted to 0;
2. centre point B3 measured and registered;
3. straight edge in position A1–C5, value of B3 from step 2 aligned and straight edge at the ends aligned to equal distances;
   the two diagonals define the plane embodiment; measurements A1 and C5 registered;
4. straight edge in position A1–C1 aligned to the already registered measured values A1 and C1, B1 measured and registered etc.

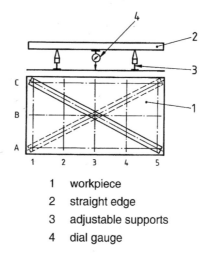

1	workpiece
2	straight edge
3	adjustable supports
4	dial gauge

Fig. 18.32 Assessment of the flatness deviation with straight edge and dial indicator

18.7.4 Assessment of roundness deviations

18.7.4.1 Definition

The deviations of the workpiece circumference from an (almost) ideal reference circle (embodiment, e.g. established by a circular movement) are measured in cross-sections perpendicular to the axis (see 18.3.2 and 18.3.3).

When the reference circle does not intersect (but eventually touches) the workpiece circumference line, the roundness deviation δ_r according to ISO 1101 is the difference between the largest and smallest radial distance of the workpiece circumference from the reference circle (Fig. 18.33 and Table 18.1).

When the reference circle does intersect the workpiece circumference line the roundness deviation δ_r according to ISO 1101 is the sum of the largest radial distances of the workpiece circumference line from the reference circle on both sides of the reference circle (Table 18.1), i.e. the range of the local deviations e of the workpiece circumference line from the reference circle (Fig. 18.34).

The reference circle is to be aligned according to the minimum requirement (Fig. 18.2). See also 18.3.2 and 18.3.3.

The roundness deviation δ_r must not exceed the roundness tolerance t_r: $\delta_r \leq t_r$.

For the assessment of the roundness deviation with coordinate measuring machines or with form measuring instruments the following reference circles are standardized according to ISO 4291:

(a) least squares circle (LSC), basis for tolerance or deviation zone Z_q;
(b) minimum zone circle (MZC), basis for tolerance or deviation zone Z_z;
(c) contacting circle (MIC, MCC), basis for tolerance or deviation zone Z_i or Z_e.

See 18.7.2.1. These reference circles intersect or touch the workpiece circumference line.

According to ISO 1101 the minimum zone circle (minimum requirement) applies. However, for practical reasons sometimes the other definitions are applied.

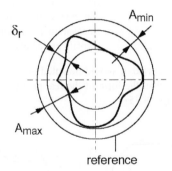

Fig. 18.33 Roundness measurement with reference circle that does not intersect the workpiece circumference line; $\delta_r = A_{max} - A_{min} \leq t_r$

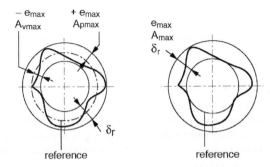

Fig. 18.34 Roundness measurement with reference circles that intersect or touch the workpiece circumference line; $\delta_r = A_{max} + A_{min} \leq t_r$

The minimum zone circle leads to the smallest values of roundness deviation. According to TGL 39 096 the (random) differences in the values of the roundness deviation caused by the different definitions of the reference circle are up to +15%.

The least squares circle needs fewer measuring points than the other reference circles to be sufficiently stable. Therefore the least squares circle is preferred in the measurement technique.

For the definition of the cross-sections perpendicular to the axis see 18.7.2.1 and 18.3.3.

A standard ISO 12 181 on the measurement of roundness deviations is in preparation.

18.7.4.2 Measuring methods

The reference circle (embodiment) may be established by:

- high precision circular movement (form measuring instrument) (measurement while workpiece mounted in centres or in a chuck);
- high precision straight guides perpendicular to each other and calculation of circles (coordinate measuring machines);
- revolving on a measuring plate (measurement of diameters);
- revolving in a V-block (three-point measurement);
- circular device (measurement in a ring or on a plug);
- circular marking (profile projector).

When polar diagrams are used, it should be noted that the form in the profile diagram looks quite different than the form of the real profile because of the amplification of radial distances (Fig. 18.35).

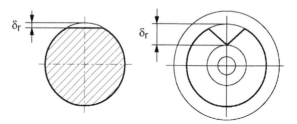

Fig. 18.35 Distortion of a profile diagram caused by the amplification in the diagram: left, without amplification and without distortion; right, with amplification and distortion

18.7.4.3 Two- and three-point measurement

For the methods, see Figs 18.36 and 18.37.

With two-point measurements (measurements of diameter), lobed forms cannot be detected (Fig. 18.36). When lobed forms can occur (e.g. with centreless grinding, reaming), in addition to the two-point measurement other inspections (e.g. three-point measurements, Fig. 18.37) are necessary.

With three-point measurements (e.g. in V-blocks, Fig. 18.37) the possibility of detecting the different types of lobed forms and other forms differs depending on the

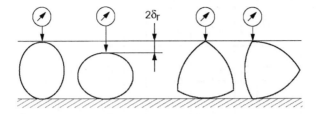

Fig. 18.36 Oval and lobed forms

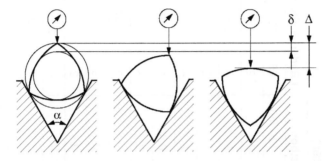

Fig. 18.37 Assessment of form deviations of lobed forms in V-blocks; summit method

type of form deviation (e.g. type of lobed form) and on the type of measuring method. Table 18.3 gives correction values k. The measured values $\Delta = A_{max} - A_{min}$ are to be divided by k in order to correspond to the roundness deviation δ_r:

$$\delta_r = \Delta/k = (A_{max} - A_{min})/k \leq t_r.$$

With three-point measurements a distinction must be made between the summit method (V-support, two fixed anvils on one side and the measuring anvil, indicator, on the other side of the workpiece; Figs 18.37 to 18.39) and the rider method (two fixed anvils and the measuring anvil, indicator, on the same side of the workpiece; Fig. 18.40).

The number of sides n of the lobed form (number of undulations) may be detected by counting the significant maximum (or minimum) indications during one revolution of the workpiece in the V-block (e.g. in the 72° V-block). When there is a superposition of harmonics (undulations) (see Figs 18.113 and 18.115), selection of the proper correction value is practically impossible. The correction value must be estimated. For this case TGL 39 096 recommends the asymmetrical summit method (Fig. 18.39) $\alpha = 120°/$ $\beta = 10°$ and the average correction value 1.2.

In order to cover all possible form deviations and numbers of undulations BS 3730 : Part 3 : 1982 recommends that one two-point measurement and two three-point measurements be taken at different angles between fixed anvils and angles be selected from the following:

symmetrical setting: $\alpha = 90°$ and $120°$ or $\alpha = 72°$ and $108°$;
asymmetrical setting: $\alpha = 120°$, $\beta = 60°$ or $\alpha = 60°$, $\beta = 30°$;

Table 18.3 Correction values k for the measurement of form deviations by two and three-point measurements: n is the order of the harmonics (number of undulations) to be assessed

Angles α and α/β of measuring methods (Figs 18.38–18.40) (TGL 39 096)

n	Summit method								I Ryder method				
	60°	72°	90°	108°	120°	180°	60°/30°	120°/60°	60°	72°	90°	108°	120°
2	–	0.47	1	1.4	1.6	2	1.4	2.4	2	1.5	1	0.62	0.42
3	3	2.6	2	1.4	1	–	2	2	3	2.6	2	1.4	1
4	–	0.38	0.41	–	0.42	2	1.4	1	2	2.4	2.4	2	1.6
5	–	1	2	2.2	2	–	2	2	–	1	2	2.2	2
6	3	2.4	1	–	–	2	0.73	0.42	1	0.38	1	2	2
7	–	0.62	–	1.4	2	–	2	2	–	0.62	–	1.4	2
8	–	1.5	2.4	1.4	0.42	2	1.4	1	2	0.47	0.4	0.62	1.6
9	3	2	–	–	1	–	2	2	3	2	–	–	1
10	–	0.70	1	2.2	1.6	2	1.4	2.4	2	2.7	1	0.24	0.42
11	–	2	2	–	–	–	–	–	–	2	2	–	–
12	3	1.5	0.41	0.38	2	2	0.73	1	1	0.47	2.4	0.62	1.6
13	–	0.62	2	1.4	–	–	–	–	–	0.62	2	1.4	–
14	–	2.4	1	–	1.6	2	1.4	0.42	2	0.38	1	2	0.42

Table 18.3 (Cont.)

Angles α and α/β of measuring methods (Figs 18.38–18.40) (TGL 39 096)

n	Summit method								I Ryder method				
	60°	72°	90°	108°	120°	180°	60°/30°	120°/60°	60°	72°	90°	108°	120°
15	3	1	–	2.2	1	–	2	2	3	1	–	2.2	1
16	–	0.38	2.4	–	0.42	2	1.4	1	2	2.4	0.41	2	1.6
17	–	2.6	–	1.4	2	–	2	2	–	2.6	–	1.4	2
18	3	0.47	1	1.4	–	2	0.73	2.4	1	1.5	1	0.62	2
19	–	–	2	–	2	–	2	2	–	–	2	–	2
20	–	2.7	0.41	2.2	0.42	2	1.4	1	2	0.7	2.4	0.24	1.6
21	3	–	2	–	1	–	2	2	3	–	2	–	1
22	–	0.47	1	1.4	1.6	2	1.4	0.42	4	1.5	1	0.62	0.42

– the method gives no indication of the roundness deviation.
180° is two-point measurement.

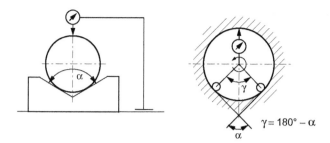

Fig. 18.38 Three-point measurement: summit method, symmetrical setting

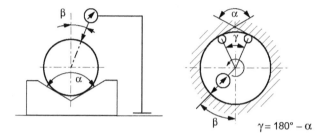

Fig. 18.39 Three-point measurement: summit method, asymmetrical setting

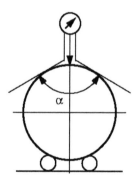

Fig. 18.40 Three-point measurement: rider method, symmetrical setting

where

α = the angle between fixed anvils

β = the angle between the direction of measurement and the bisector of the angle between fixed anvils (Fig. 18.39)

The measuring anvil should be selected from Table 18.4. ISO 4292 gives a similar table, but does not specify spherical or cylindrical anvils.

Table 18.4 Types of anvil (BS 3730: Part 3: 1982)

Surface form	Anvil radius in mm	Surface radius in mm
Convex surface	Spherical: 2.5	All
Convex edge	Cylindrical: 2.5	All
Concave surface	Spherical: 2.5	≥ 10
Concave edge	Cylindrical: 2.5	≥ 10
Concave surface	Spherical: 0.5	<10
Concave edge	Cylindrical: 0.5	<10

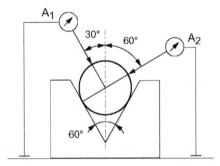

Fig. 18.41 Combination of two- and three-point measurement according to U. Barth for the determination of the smallest actual size, the mating size and the roundness deviation

It is further recommended that the following fixed anvils be used:

- for external measurement: V-support with a small radius; the median plane of the V-support should be in the same plane as the plane of measurement;
- for internal measurement: sphere with a small radius; the median plane of the sphere should be in the same plane as the plane of measurement.

With a combination of two-point measurement, three-point measurement and evaluation by calculation according to U. Barth (see Ref. [10]) (Fig. 18.41), approximations of the smallest diameter, the mating size and the roundness deviation can be assessed. This method can assess form deviations up to the 10th harmonic (e.g. number of sides of the lobed form). It is not necessary to know the number of the harmonic.

The actual sizes are detected by the dial indicator A_2. The mating size P and the roundness deviation δ_r* are to be calculated as follows:

When the indicator reading differences $\Delta A_1 \leqslant A_2$:

$$P \approx A_{2\text{min}} + \Delta A_2 = A_{2\text{max}}$$
$$\Delta_r \approx \Delta A_2/2 \leqslant t_r$$

* Applicable to disk-shaped workpieces or when the straightness and parallelism deviations are negligible.

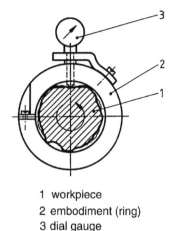

1 workpiece
2 embodiment (ring)
3 dial gauge

Fig. 18.42 Assessment of roundness deviations with a measuring ring

When the indicator reading differences $\Delta A_1 > \Delta A_2$:

$$P \approx A_{2min} + \Delta A_1/2$$
$$\delta_r \approx \Delta A_1/2 \leq t_r$$

This method is relatively precise for short (disk-shaped) workpieces (which are practically not bent along the axis).

18.7.4.4 Assessment with measuring ring or measuring plug

Figure 18.42 shows a measuring ring according to the former East German Standard TGL 39 096. The inner ring surface establishes the circle embodiment. The diameter is adjustable within the range of the size tolerance.

For the assessment of the roundness deviations of holes a similar method with measuring plugs can be used.

These assessments are close to the definition of the roundness deviation according to ISO 1101.

18.7.5 Assessment of cylindricity deviations

18.7.5.1 Definition

The deviations of the workpiece surface from an almost ideal reference cylinder (embodiment, e.g. established by circular movements and movements perpendicular to the plane of the circular movements) are measured.

When the reference cylinder does not intersect the workpiece surface (but eventually touches it), the cylindricity deviation δ_z according to ISO 1101 is the difference between the largest and smallest radial distances of the workpiece surface from the reference cylinder. Similarly to Fig. 18.33, it is

$$\delta_z = A_{max} - A_{min}.$$

When the reference cylinder does intersect the workpiece surface, the cylindricity deviation according to ISO 1101 is the sum of the largest radial distances of the workpiece surface from the reference cylinder on both sides of the reference cylinder (Table 18.1), i.e. the range of the local deviations e of the workpiece surface from the reference cylinder (Fig. 18.34).

The reference cylinder is to be aligned according to the minimum requirement (Fig. 18.2). When methods (a), (b), (c) and (e) of 18.7.5.2 (measurement in sections) are used, the alignment of the workpiece according to 18.3.2 must be observed (see also 18.3.3).

The cylindricity deviation δ_z must not exceed the cylindricity tolerance t_z: $\delta_z \leq t_z$.

For the assessment of the cylindricity deviation with coordinate measuring machines or with form measuring instruments the following reference cylinders are used:

(a) least squares cylinder (according to Gauss, the cylinder of which the sum of the squares of the radial distances to the surface is a minimum, regression cylinder);
(b) minimum zone cylinder (according to Chebyshev, the cylinder of which the maximum radial distance to the surface is a minimum);
(c) contacting cylinder (the maximum inscribed cylinder for holes, minimum circumscribed cylinder for shafts).

These reference cylinders intersect or touch the workpiece surface. According to ISO 1101 the minimum zone cylinder (minimum requirement) applies. However, for practical reasons sometimes the other definitions are applied.

The minimum zone cylinder leads to the smallest values of cylindricity deviation δ_z.

The least squares cylinder needs fewer measuring points than the other reference cylinders to be sufficiently stable. Therefore the least squares cylinder is preferred in the measurement technique.

A standard (ISO 12 180) on the measurement of cylindricity deviation is in preparation.

18.7.5.2 Measuring strategies

The following measuring strategies are to be distinguished:

(a) radial section method;
(b) generatrix method;
(c) generatrix and radial section method;
(d) helical method;
(e) extreme positions method;
(f) point method.

In all methods the measured profiles or points must be related to the same coordinate system. The workpiece (feature) axis has to be aligned parallel to the straight line guidance of the measuring device. This alignment can also be replaced by calculation. For the definition of the axis see 18.7.2.1. See also 18.3.3.

(a) Radial section method

The profile lines (circumferences) of several cross-sections perpendicular to the reference cylinder (axis of measurement) are to be plotted in one common polar diagram and evaluated according to the minimum requirement (Fig. 18.43).

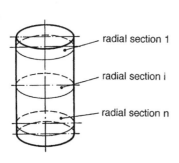

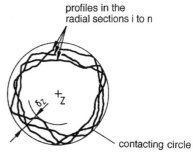

profiles in the
radial sections i to n

radial section 1

radial section i

radial section n

contacting circle

Z centre of contacting circle

Fig. 18.43 Radial section method

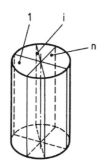

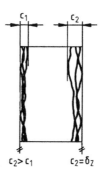

Fig. 18.44 Generatrix method

(b) Generatrix method

In several sections containing the reference cylinder axis (axis of measurement) the two always opposite profile lines (generatrixes) are assessed, plotted in one common diagram and evaluated according to the minimum requirement (Fig. 18.44).

(c) Generatrix and radial section method

This is a combination of the generatrix method and the radial section method. It gives the smallest measurement uncertainty because in both directions (axial and radial) the deviations are assessed with small sampling intervals (e.g. according to the Nyquist theorem) (see 18.9.1). Therefore this method is recommended.

(d) Helical method

The profile lines assessed by a helical probing trace, probing perpendicular to the reference cylinder axis (axis of measurement), are plotted in one common diagram and evaluated according to the minimum requirement (Fig. 18.45). Eventually the profiles of two cross-sections at the features end perpendicular to the reference cylinder (axis of measurement) will also be assessed and included in the diagram and evaluation.

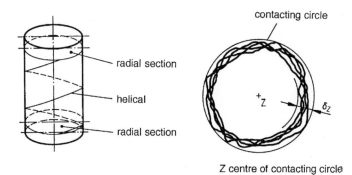

Fig. 18.45 Helical method

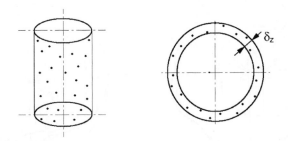

Fig. 18.46 Point method

(e) Point method

Random distributed points of the workpiece surface are assessed, plotted in one common polar diagram perpendicular to the reference cylinder (axis) and evaluated according to the minimum requirement (Fig. 18.46).

(f) Extreme positions method

According to TGL 39 097, the following are assessed and plotted:

- generatrix in one section containing the reference cylinder axis (generator lines at 0° and 180°);
- circumference lines in cross-sections perpendicular to the reference cylinder axis in the positions where the largest (max) and the smallest (min) distances of the generator lines occur (cross-sections I and II).

The evaluation is shown in Fig. 18.47. Z_1, c_{max} and Z_2, c_{min} in the section 0°–180° are the same in both Fig. 18.47 top and bottom. The cylindricity deviation is $\delta_z = a_1 + a_2 + a_3 \leqslant t_z$.

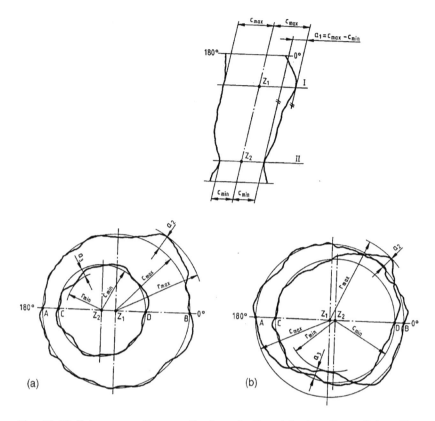

Fig. 18.47 Extreme positions method, evaluation: (a) when the radial profiles do not intersect; (b) when the radial profiles intersect

18.7.5.3 Measuring methods

18.7.5.3.1 Measuring the cylindricity deviation

The reference cylinder (embodiment) may be established by:

• high precision circular movement and high precision straight guides perpendicular to the circular movement	(form measuring instrument) (measuring while workpiece is mounted in centres or in a chuck);
• high precision straight guides perpendicular to each other and cylinder calculation	(coordinate measuring machine);
• revolving in V-block and parallel straight guides	(three-point measurement with V-block, straight edge, dial indicator and measuring table).

The three-point measurement does not allow establishing a stable workpiece coordinate system (the workpiece axis has no stable position in the coordinate system of measurement). Therefore this method is a rough approximation (see 18.7.4.3).

18.7.5.3.2   Assessment by measurement of components

(a) Approximate evaluation of the cylindricity deviation δ_z from the roundness deviation δ_r and the parallelism deviation δ_p* of the generator lines:

$$\delta_z \approx \delta_r + \delta_p \leq t_z$$

(b) Approximate evaluation of the cylindricity deviation δ_z from the roundness deviation δ_r and the longitudinal section profile deviation δ_q (see 21.4):

$$\delta_z \approx \delta_r + \delta_q \leq t_z$$

The approximation (b) with δ_q is considered to be closer to the correct assessment of the cylindricity deviation than the approximation (a) (with δ_p) (see Ref. [12]).

(c) Approximate evaluation of the cylindricity deviation δ_z from the total run-out deviation δ_t relative to another cylindrical feature:

$$\delta_z \approx \delta_t \leq t_z$$

The total run-out deviation δ_t is always larger than the cylindricity deviation δ_z because δ_z comprises the cylindricity deviation and the coaxiality deviation.

18.7.6 Assessment of profile deviations of lines

18.7.6.1 Definition

The deviations of the workpiece section (profile) line from an (almost) ideal reference line (embodiment) are measured in cross-sections, the orientation of which is defined by the datum system (related profile tolerance) or parallel to the drawing projection plane (unrelated profile tolerance) or for surfaces of revolution containing the workpiece established axis (of another feature) as described in 18.3.3.1.

The location and orientation of the reference line are defined relative to the datum system (related profile tolerance) as explained in 18.7.9 or are defined by the minimum requirement, i.e. the maximum distance of the surface line from the reference line shall be a minimum (unrelated profile tolerance).

(Regarding the alignment of the workpiece established axis, in principle the similar effects occur as explained in 18.3.3.2 and 18.3.3.3.)

The profile deviation δ_b is the maximum distance of the workpiece section (profile) line from the reference line, measured perpendicular to the reference line (Table 18.1).

The profile deviation δ_b must not exceed half of the profile tolerance t_b: $\delta_b \leq t_b/2$.

---

* The cylindricity deviation comprises deviations of roundness, straightness and parallelism. As the parallelism deviation δ_p according to ISO 1101 comprises the deviations and the inclinations of the generator lines relative to each other, the cylindricity deviation δ_z cannot be larger than the sum of the roundness deviation δ_r and the parallelism deviation δ_p. In most cases it is smaller (see Ref. [12]).

18.7.6.2 Measuring methods

The reference line (embodiment) may be established by:
- profile template   (copying system*)
- profile marking   (profile projector)
- high precision straight guides   (coordinate measuring machine)
  perpendicular to each other
  and calculation of the reference line

18.7.7 Assessment of profile deviations of surfaces

18.7.7.1 Definition

The deviations of the workpiece surface from an (almost) ideal reference surface (embodiment, e.g. established by a form template or by calculation in a coordinate measuring machine) are measured.

For the location and orientation of the reference element (alignment of the workpiece) the same applies as to the assessment of profile deviations of lines (see 18.7.6.1).

The profile deviation δ_h is the maximum distance of the workpiece surface from the reference surface, measured perpendicular to the reference surface (Table 18.1).

The profile deviation δ_h must not exceed half of the profile tolerance t_h: $\delta_h \leq t_h/2$.

18.7.7.2 Measuring methods

The reference line (embodiment) may be established by:

- surface template   (copying system*)
- high precision straight guides   (coordinate measuring machine)
  perpendicular to each other
  and calculation of the reference line
- profile template
  (for surfaces of revolution only)   (rotating profile template device)

18.7.8 Assessment of orientation deviations

18.7.8.1 Definition

The deviations of the workpiece surface or the workpiece line (axis, generator line) from an (almost) geometrical ideal reference element (plane or straight line embodiment) are measured. The reference element is to be aligned according to the datum or datum system (parallel, perpendicular, in the specified angle). The orientation of the datum is determined by the minimum rock requirement at the datum element of the workpiece (see 18.3.2) or by the adjustment of the workpiece in the datum system (see 3.4).

When the reference element does not intersect the workpiece feature (but eventually touches it) the orientation deviation δ_d according to ISO 1101 is the difference between the largest and smallest distances of the workpiece feature from the reference element (Table 18.1).

---

* The copying system probes the workpiece and the template with the same tip radius and records the deviations of the workpiece from the template.

When the reference element does intersect the workpiece feature the orientation deviation δ_d according to ISO 1101 is the sum of the largest distances of the workpiece feature from the reference element on both sides of the reference element (Table 18.1).

The orientation deviation δ_d must not exceed the orientation tolerance t_d: $\delta_d \leq t_d$.

Note that this definition is in accordance with ISO 1101 and ISO 5459. The orientation deviation encloses the flatness deviation of the surface to be measured or the straightness deviation of the axis or line to be measured.

According to the former Comecon standard ST RGW 301-76, when measuring the orientation deviation, the form deviation of the feature to be measured is to be eliminated, e.g. by using a parallel plane plate or a mandrel contacting the workpiece surface and taking the measurements from this auxiliary element (plate, mandrel).

With the definition of the orientation tolerance in ISO 1101, which includes the form deviation of the toleranced feature, it was assumed that normally no larger form deviations are permitted by the function than the orientation tolerance. This enables simple inspection (measurement).

18.7.8.2 Combination of features

There are possible deviations of parallelism, perpendicularity or angularity:

- of a planar surface relative to a planar surface;
- of a planar surface relative to a straight line (axis, generator line);
- of a straight line (axis, generator line) relative to a planar surface;
- of a straight line (axis, generator line) relative to a straight line (axis, generator line).

18.7.8.3 Measurement methods

18.7.8.3.1 General

For the reference straight line (embodiment of straight line) see 18.7.1.3 and for the reference plane (embodiment of a plane) see 18.7.3.2.

The reference angle (embodiment of angle) may be established by:

- high precision divided circle   (dividing head);
- high precision straightness guides,   (form measuring instrument)
  perpendicular to each other   (coordinate measuring machine);
- solid angle   (measuring plate with measuring angle)
                                         (sine bar rule);
- optical beam   (pentaprism).

With the use of mandrels, the cylindricity deviation of the workpiece hole to be measured is eliminated from the measurement. Therefore this method is an approximation according to ISO 1101 (but a method close to the definition according to the former Comecon Standard ST RGW 301-76; see Fig. 18.14.

18.7.8.3.2 Assessment of orientation deviations of straight generator lines or planar surfaces by measuring distances

For the measuring method see Fig. 18.48.

The orientation deviation δ_d (deviation of parallelism, deviation of perpendicularity) is the difference between the largest distance A_{max} and the smallest distance A_{min} (indicator

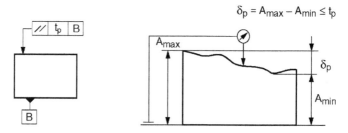

Fig. 18.48 Assessment of the orientation deviation δ_d (here parallelism deviation δ_p) of a surface or generator line

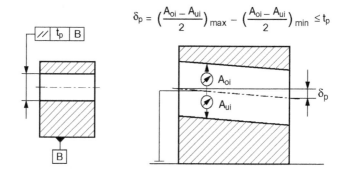

Fig. 18.49 Assessment of the parallelism deviation δ_d of an axis, evaluated from measurement of distances

reading) of the workpiece feature from the measuring table or from the solid angle, and must not exceed the orientation tolerance t_d: $\delta_d \leq t_d$.

18.7.8.3.3 Assessment of the orientation deviation of an axis by measuring distances

For the measuring method see Fig. 18.49.

The differences of the distances from the measuring plate are to be measured. During the measurements, the indicators must keep their adjustment, but need not be calibrated in the height.

As the measuring directions are opposite, the difference of half of the sums of the distances from the measuring plate is assessed.

Figure 18.50 shows the assessment of the orientation deviation with two parallel measuring plates. Figure 18.51 shows the assessment by distance and diameter measurements.

When the tolerance zone is cylindrical, the orientation deviations are to be assessed in two mutually perpendicular axial sections. The cylindrical orientation deviation in this case is

$$\delta_d = \sqrt{(\delta_{dx}^2 + \delta_{dy}^2)}$$

(see 18.6).

$$\delta_n = \sqrt{(\delta_{nx}^2 + \delta_{ny}^2)} \leq t_n \qquad \delta_{nx} \approx \frac{(A_{ro} - A_{ru}) - (A_{lo} - A_{lu})}{2} \frac{l}{l_m} \leq t_n$$

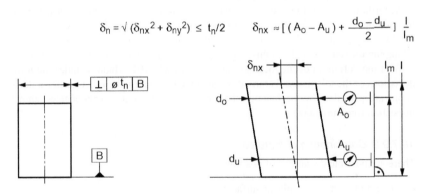

Fig. 18.50 Assessment of the perpendicularity deviation δ_n of an axis, evaluated from measurements of distances from parallel planes

$$\delta_n = \sqrt{(\delta_{nx}^2 + \delta_{ny}^2)} \leq t_n/2 \qquad \delta_{nx} \approx [(A_o - A_u) + \frac{d_o - d_u}{2}] \frac{l}{l_m}$$

Fig. 18.51 Assessment of the perpendicularity deviation δ_n of an axis, evaluated from measurements of distances and diameters

18.7.8.3.4 Assessment of orientation deviations with form measuring instruments or coordinate measuring machines

For the assessment of the orientation deviation of a planar surface or a straight generator line see 18.7.8.3.2.

The orientation deviation of an axis can be assessed by measurements of distances as described in 18.7.8.3.3 or by the centres of cross-sections assessed from the circumference lines in cross-sections perpendicular to the axis.

For the assessment of cross-section centres see 18.7.2.1.

18.7.9 Assessment of location deviations

18.7.9.1 Definition

The deviations of the workpiece surface (planar surface) or the workpiece line (axis) from an (almost) ideal reference element (plane or straight line embodiment) are measured. The reference element (embodiment) is to be aligned according to the datum or datum system (parallel, perpendicular, in the specified angle). The orientation of the datum is determined by the minimum rock requirement at the datum element of the workpiece (see 18.3.2) or by the adjustment of the workpiece in the datum system (see 3.4). The distance between the reference element and the datum (theoretical exact location of the embodiment) is zero (coaxiality, symmetry) or specified by theoretical exact dimensions (in rectangular frames). The same applies to the distances of the reference elements between each other, if applicable.

The embodiment can be located apart from the theoretical exact location, but then the measured values have to be corrected accordingly.

The location deviation δ_o according to ISO 1101 is the largest distance of the workpiece feature (surface, axis) from the reference element (which is or is considered to be in the theoretical exact location and orientation) (Figs 18.52 and 18.53 and Table 18.1), and must not exceed half of the location tolerance t_o: $\delta_o \leq t_o/2$.

Note that this definition is in accordance with ISO 1101 and ISO 5459. The location deviation encloses the flatness deviation of the surface to be measured or the straightness deviation of the axis to be measured.

According to the former Comecon standard ST RGW 301-76, when measuring the location deviation, the form deviation of the feature to be measured is to be eliminated, e.g. by using a parallel plane plate or a mandrel contacting the workpiece surfaces and taking the measurements from this auxiliary element (plate, mandrel) (Fig. 18.14).

With the definition of the location tolerance in ISO 1101, which includes the form deviations of the toleranced feature, it was assumed that normally no larger form deviations are permitted by the function than the location tolerance. This enables simple inspection (measurement).

18.7.9.2 Combination of features

There are possible deviations of position:

- of a point relative to a planar surface;
- of a point relative to a straight line (axis, generator line);
- of a point relative to a point;
- of a straight line (axis, generator line) relative to a planar surface;
- of a straight line (axis, generator line) relative to a straight line;
- of a planar surface relative to a planar surface;
- of a surface relative to a datum system;
- of a line relative to a datum system;
- of a point relative to a datum system.

There are possible deviations of coaxiality:

- of a straight line (axis) relative to a straight line (axis);
- of a centre point relative to a straight line (axis).

There are possible deviations of symmetry:

- of a straight line (axis) relative to a planar surface (symmetry face*);
- of a planar face (symmetry face) relative to a planar face (symmetry face*).

* Also referred to as median face.

18.7.9.3 Measuring methods

18.7.9.3.1 General

For the reference element (embodiment of straight line, of plane) see 18.7.1.3. For the embodiment of the orientation see 18.7.8.3.1.

The embodiments are to be located in the theoretical exact distance (zero or specified theoretical exact dimension in rectangular frame) from the datum, or the measured values are to be corrected accordingly.

18.7.9.3.2 Assessment of the location deviation of a straight generator line or a planar surface by measuring distances

For the measuring method see Fig. 18.52.

The location deviation δ_o is the absolute value of the maximum difference between the measured distance A and the theoretical exact distance A_{th} and must not exceed half of the location tolerance t_o: $\delta_o \leq t_o/2$.

$$\delta_o = |A - A_{th}|_{max} \leq t_o/2$$

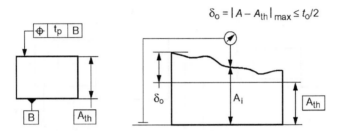

Fig. 18.52 Assessment of location deviation of a generator line or a surface

18.7.9.3.3 Assessment of the location deviation of an axis relative to plane surface by measuring distances

For the measuring method see Fig. 18.53.

Both dial indicators are calibrated to the distance from the measuring plate. In each section the arithmetical mean of the distances from the measuring plate $(A_{oi} + A_{ui})/2$ is

$$\delta_o = \left| \frac{A_{oi} + A_{ui}}{2} - A_{th} \right|_{max} \leq t_o/2$$

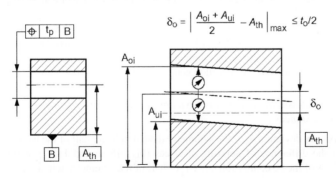

Fig. 18.53 Location deviation δ_o of the axis of the hole calculated from distance measurements

assessed. The location deviation δ_o is the largest absolute value of the difference between the arithmetical mean and the theoretical exact distance, and must not exceed half of the location tolerance t_o: $\delta_o \leqslant t_o/2$.

When the tolerance zone is cylindrical, the values $A_i = (A_{oi} + A_{ui})/2 - A_{th}$ (Fig. 18.54) must be assessed in each axial location in sections perpendicular to each other, A_{ix} and A_{iy}. The cylindrical location deviation δ_o is the largest value of $\sqrt{(A_{ix}^2 + A_{iy}^2)}$ and must not exceed half of the location tolerance t_o: $\delta_o \leqslant t_o/2$.

When the theoretical exact distance from the datum is specified by theoretical exact dimensions (rectangular framed) arranged in a chain, the sum of these dimensions applies to the distance of the tolerance zone and reference element from the datum (Fig. 18.55).

$$A_i = \frac{A_{oi} + A_{ui}}{2} - A_{th}$$

local location deviation $\delta_o = \sqrt{(A_{ix}^2 + A_{iy}^2)} \leq t_o/2$

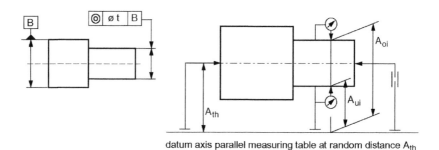

datum axis parallel measuring table at random distance A_{th}

Fig. 18.54 Assessment of the location deviation δ_o (here coaxiality deviation δ_a) of a shaft by measuring distances

18.7.9.3.4 Assessment of location deviations with form measuring instruments or coordinate measuring machines

For the assessment of the location deviation of a planar surface or a straight generator line see 18.7.9.3.2.

The location deviation of an axis can be assessed by measurements of distances as described in 18.7.9.3.3 or by the centres of cross-sections assessed from the circumference lines in cross-sections perpendicular to the axis.

For the assessment of cross-section centres see 18.7.2.1.

18.7.9.3.5 Assessment of coaxiality deviations or symmetry deviations by measuring distances

The coaxiality deviation δ_a or the symmetry deviation δ_s are half of the largest absolute value of the difference of opposite distances (Figs 18.56 and 18.57), and must not exceed half of the coaxiality tolerance t_a or half of the symmetry tolerance t_s:

$$|A_{i1} - A_{i2}|_{max}/2 \leqslant t_a/2$$

$$|A_{i1} - A_{i2}|_{max}/2 \leqslant t_s/2$$

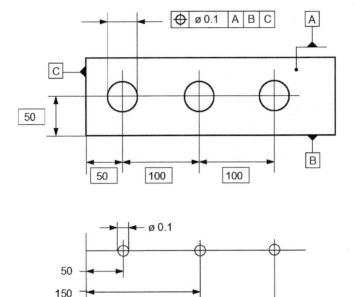

Fig. 18.55 Addition of theoretical exact dimensions arranged in a chain

$$\delta_a = \frac{|\,A_{i1} - A_{i2}\,|_{max}}{2} \leq t/2$$

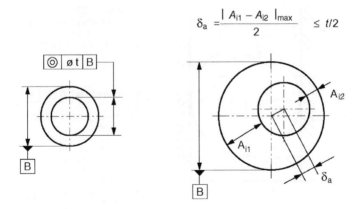

Fig. 18.56 Assessment of the coaxiality deviation δ_a by measuring distances

Figure 18.58 shows the inspection of a keyway for a key assembly with interference fits at the key and at the shaft and hub (see also 20.12).

18.7.9.3.6 Inspection of coaxiality deviations by measuring run-out deviations

Measuring the coaxiality deviation is often complicated and costly. Therefore, for the inspection of the workpiece often the more simple run-out deviations are assessed

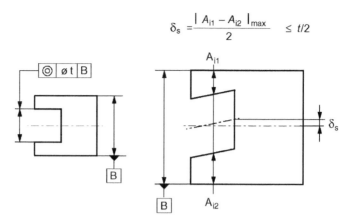

Fig. 18.57 Assessment of the symmetry deviation δ_s by measuring distances

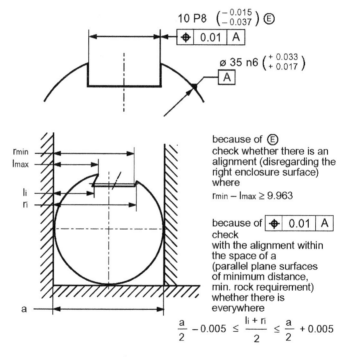

Fig. 18.58 Measurement of the symmetry deviation of a keyway (toleranced regardless of feature size)

(see 18.7.10.2.1) and compared with the value of the coaxiality tolerance. Only when the run-out deviation exceeds the coaxiality tolerance is it checked whether twice of the coaxiality deviation exceeds the coaxiality tolerance (or whether the larger run-out deviation is caused by roundness deviations only, which is permissible).

18.7.9.3.7 Assessment of coaxiality deviations or symmetry deviations with form measuring instruments or coordinate measuring machines

The coaxiality deviations or the symmetry deviations can be assessed by measurements of opposite distances as described in 18.7.9.3.5.

The coaxiality deviation can also be assessed by the centres of cross-sections assessed by the circumference lines in cross-sections perpendicular to the axis. For the assessment of cross-sections centres see 18.7.2.1 and Fig. 8.26.

For the definition of symmetry faces* (median face) see Fig. 8.25.

18.7.10 Assessment of run-out deviations

18.7.10.1 Datums

For the assessment of the run-out deviation the workpiece is to be aligned according to the datum axis, or the coordinates are to be transformed accordingly by calculation.

For the datum the minimum rock requirement according to ISO 5459 applies (see 18.3.2).

The following practical solutions are available.

(a) Mandrel
    The mandrel should fit into the hole without clearance. If the mandrel rocks, the minimum rock requirement applies.

(b) Centres
    This is according to the definition when the drawing indicates the centre holes as datum. When the drawing indicates the axis of a cylindrical feature (not the centre holes) as the datum, the measurement is made incorrect by the form deviations and the eccentricity of the centre holes relative to the correct datum axis (of the cylindrical feature).

(c) Chuck
    The chuck must have a small run-out deviation in comparison with the run-out tolerance. The suitability can be checked prior to the measurement by measuring the run-out deviation of an almost perfect cylinder embodiment in the chuck. If necessary and possible, the workpiece datum feature is to be aligned within the chuck with the aid of an indicator to indicate the least possible indication difference.

(d) Revolving table
    The datum feature is to be aligned with the aid of an indicator in two cross-sections, (e.g. 1/8 of the datum feature length apart from the ends) according to the least possible indication difference. This alignment deviates from the theoretical exact datum according to ISO 5459 when the datum feature axis deviates from straightness.

(e) Coordinate measuring machine
    From the assessed (probed) points of the datum feature surface the contacting cylinder (see 18.7.2.1) must be determined. When the contacting cylinder can rock, the minimum rock requirement according to ISO 5459 applies (see 18.3.2). However, practically the contacting cylinder is aligned parallel to the regression cylinder (because the appropriate mathematical technique has been developed).

---

* Similar to the way in which the actual axis deviates from a straight line according to the form deviations of the cylindrical surface, the actual symmetry face deviates from a plane according to the form deviations of the planar surfaces establishing the feature.

(f)   Form measuring instrument
      Depending on the type of the instrument procedure, (d) or (e) applies.
(g)   V-blocks, measuring table
      This method should be applied only when the cylindricity deviation of the datum
      feature is small in comparison with the run-out tolerance. In other cases the mea-
      surement is made considerably inaccurate by the form deviations (see 18.7.4.3).

18.7.10.2 Measuring methods of circular run-out

The deviations of the workpiece surface section line from an (almost) geometrical ideal
reference circle (circle embodiment) coaxial (concentrical) with the datum axis are
measured in sections which are plane and perpendicular to the datum, or conical or
cylindrical and coaxial to the datum. The circle embodiments are established by revolving the workpiece (Fig. 18.59) or the indicator (Fig. 18.60) using:

- a length measuring instrument and workpiece support by
  - a mandrel between centres,
  - centres,
  - a chuck,
  - a revolving table;
- a form measuring instrument.

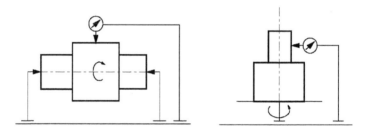

Fig. 18.59 Measurement of the run-out deviation with revolving workpiece

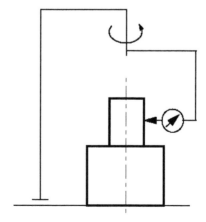

Fig. 18.60 Measurement of the run-out deviation with revolving indicator

In coordinate measuring machines the circle embodiment is established by calculation relative to the almost straight and perpendicular axes of the coordinate measuring machine.

18.7.10.2.1 Assessment of circular radial run-out deviations

The deviations of the workpiece circumference line of cylindrical features in sections perpendicular to the datum axis from an (almost) geometrical ideal reference circle (circle embodiment, simulated circle, ISO 5459) are measured. The circle embodiment is coaxial (concentrical) to the datum axis (Fig. 18.61).

When the reference circle does not intersect the workpiece circumference line, the circular radial run-out deviation δ_l is the difference between the largest, A_{max}, and the smallest, A_{min}, distances of the workpiece circumference line from the reference circle, and must not exceed the run-out tolerance t_l. Using dial indicators, this is the difference between the largest and smallest indication during one revolution: $\delta_l = A_{max} - A_{min} \leq t_l$.

Each section is to be considered independently from the others.

The radial run-out tolerance and deviation is only applicable to cylindrical or sectors of cylindrical features relative to cylindrical datum features.

Note that the circular radial run-out deviation comprises the eccentricity and parts of the roundness deviation of the feature to be measured. When (theoretically) the roundness deviation is zero, the circular radial run-out deviation is twice the eccentricity (Fig. 18.62).

When the workpiece feature to be measured is not very short (so that one measurement is not sufficient), the circular radial run-out deviation should be measured at least in the centre and near the ends of the workpiece feature.

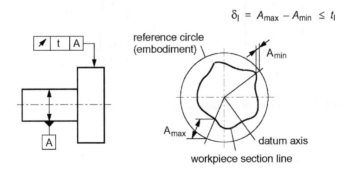

$$\delta_l = A_{max} - A_{min} \leq t_l$$

Fig. 18.61 Assessment of circular radial run-out deviation

18.7.10.2.2 Assessment of circular axial run-out deviations

The deviations of (circular) section lines of the (planar) workpiece surface in cylindrical sections coaxial with the datum axis (measuring cylinders) from an (almost) geometrical ideal reference plane are measured. (Fig. 18.63).

When the reference plane does not intersect the workpiece section line, the circular axial run-out deviation δ_l is the difference between the largest, A_{max}, and smallest, A_{min},

$$\delta_\mathrm{l} = A \leq t_\mathrm{l} \qquad \text{eccentricity} = A/2$$

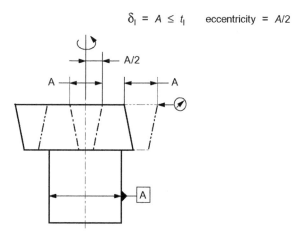

Fig. 18.62 Assessment of circular radial run-out deviation

$$\delta_\mathrm{l} = A_{max} - A_{min} \leq t_\mathrm{l}$$

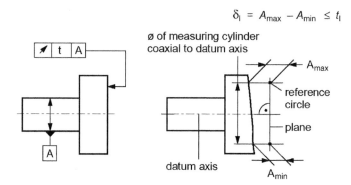

Fig. 18.63 Assessment of circular axial run-out deviation

distances of the workpiece section line from the reference plane (reference circle within the reference plane concentric to the datum), and must not exceed the run-out tolerance t_l. Using dial indicators this is the difference between the largest and smallest indications during one revolution: $\delta_\mathrm{l} = A_{max} - A_{min} \leq t_\mathrm{l}$.

Each cylindrical section is to be considered independently from the others.

Note that the circular axial run-out deviation is equal to the perpendicularity deviation of the circular section line of the surface, but may be smaller than the flatness deviation and the perpendicularity deviation of the entire surface (Fig. 3.8).

The circular axial run-out deviation should be measured at ≈ 1, 0.75 and 0.5 times the outer diameter. With surfaces manufactured by metal removing, the measurement near the outer diameter is in general sufficient, because here the largest run-out deviation occurs.

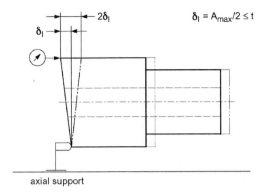

axial support

Fig. 18.64 Assessment of the circular axial run-out deviation δ_l with axial support at the outer diameter of the surface to be measured

During measurement of the circular axial run-out deviation, the workpiece and indicator must be fixed in the axial direction by using

- a chuck,
- an axial support against an auxiliary datum surface that is plane, perpendicular to the axis and not less in diameter than the measured surface,
- an axial support coaxial to the datum axis,
- a support at the surface to be measured (Fig. 18.64).

Support at a surface different from the surface to be measured, where the support is apart from the datum axis, should be avoided because deviations of form and orientation of the surface invalidate the measuring result.

When the support is at the surface to be measured and near the outer diameter, twice the value of the circular axial run-out deviation will be indicated (Fig. 18.64).

18.7.10.2.3 Assessment of circular run-out deviations in any or in a specified direction

The deviations of (circular) section lines of the workpiece surface (of rotationally symmetric features) in conical sections coaxial with the datum axis from an (almost) geometrical ideal reference circle (circle embodiment) within the conical section and coaxial with the datum are measured. If not otherwise specified, the conical section (measuring cone, measuring direction) is perpendicular to the surface to be measured (Fig. 18.65, 3.25).

When the reference circle does not intersect the workpiece section line, the circular run-out deviation (in any direction or in the specified direction) δ_1 is the difference between the largest, A_{max}, and smallest, A_{min}, distances of the workpiece section line from the reference circle, and must not exceed the run-out tolerance t_1.

Using dial indicators, this is the difference between the largest and smallest indication during one revolution: $\delta_1 = A_{max} - A_{min} \leq t_1$.

Each conical section is to be considered independently from the others.

When the workpiece feature to be measured is not very short (so that one measurement is not sufficient), the circular run-out deviation (in any direction or in the specified direction) should be measured in the centre and near the ends of the workpiece feature.

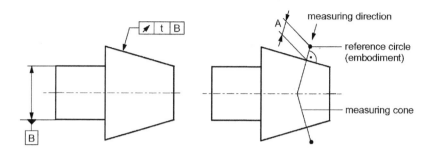

Fig. 18.65 Assessment of circular run-out deviation δ_l in any direction

For axial support of the workpiece the same applies as with the circular axial run-out deviation (see 18.7.10.2.2).

Note that the circular run-out deviation in any or in the specified direction is composed of the eccentricity and parts of the roundness deviations of the workpiece feature to be measured.

18.7.10.3 Assessment of total run-out deviations

The total run-out deviations

- total radial run-out deviation
- total axial run-out deviation

are to be distinguished from the circular run-out deviations. With total run-out deviations, the measuring sections are to be considered as dependent on each other. The circle embodiments establish the following:

- with the assessment of the total radial run-out deviation, one reference cylinder coaxial with the datum axis (i.e. using length measuring instruments, the instrument is to be guided along a straight line parallel to the datum axis) (Fig. 18.66);
- with the assessment of the total axial run-out deviation, one reference plane perpendicular to the datum axis (i.e. using length measuring instruments, the instrument is to be guided along a straight line perpendicular to the datum axis) (Fig. 18.67).

When the reference element (cylinder, plane) does not intersect the workpiece surface, the total run-out deviation δ_t is the difference between the largest, A_{max}, and smallest, A_{min}, distances of the workpiece surface from the reference element, and must not exceed the total run-out tolerance t_t. Using dial indicators, this is the difference between the largest and smallest indications during several (all) revolutions along the workpiece feature: $\delta_t = A_{max} - A_{min} \leq t_t$.

$$\delta_t = A_{max} - A_{min} \leq t_t$$

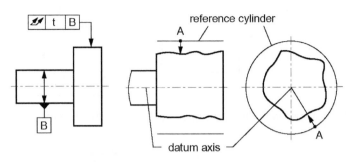

Fig. 18.66 Assessment of the total radial run-out deviation

$$\delta_t = A_{max} - A_{min} \leq t_t$$

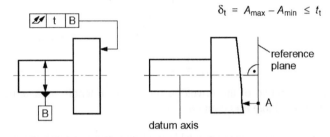

Fig. 18.67 Assessment of the total axial run-out deviation

18.7.11 Inspection of the envelope requirement

18.7.11.1 Definition

It is to be checked whether the surface of the workpiece feature (cylindrical surface or two parallel opposite plane surfaces) is contained within the (almost) geometrical ideal envelope of maximum material size (embodiment, reference element).

18.7.11.2 Inspection methods

The embodiment of the envelope of maximum material size may be established by:

- functional gauge (gauge covering the entire length of the feature, e.g. plug, ring, squared block out of gauge blocks);
- calculation of the rectangular coordinates related to the straight and perpendicular guides of the coordinate measuring machine (simulated gauge);
- calculation of the cylinder coordinates related to the straight and perpendicular guides of the form measuring instrument (simulated gauge);
- revolutions on a revolving table, (solid) straight and perpendicular guides and length measuring instrument;
- revolutions between centres, measuring table and length measuring instrument;
- revolutions in a chuck, measuring table and length measuring instrument;

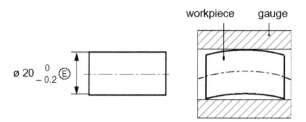

Fig. 18.68 Inspection of the envelope requirement with a functional gauge

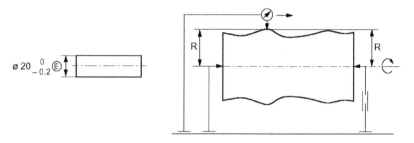

Fig. 18.69 Inspection of the envelope requirement: length measuring instrument, measuring table and centres

- revolutions in V-block, measuring table and length measuring instrument;
- measuring table and length measuring instrument (for features composed of two parallel opposite plane surfaces).

18.7.11.2.1 Inspection with functional gauge

When the gauge covers the entire surface of the workpiece feature, this is an inspection close to the definition (Fig. 18.68).

18.7.11.2.2 Inspection with coordinate measuring machine

When computer programs are applied to simulate the functional gauge, and sufficient points of the feature's surface are assessed (probed), this is an inspection close to the definition.

18.7.11.2.3 Inspection with form measuring instrument

When computer programs are applied to simulate the functional gauge, and sufficient sections of the workpiece feature are assessed, this is an inspection close to the definition.

18.7.11.2.4 Inspection with length measuring instrument and measuring table, centres or chuck or revolving table

The inspection method is shown in Fig. 18.69.

The reference cylinder is established by the axis of revolution parallel to the measuring table (e.g. established by centres one of which is adjustable) and a length measuring instrument calibrated to a distance (radius) from the axis of revolution of half of the maximum material size (e.g. calibrated with a disc of known diameter).

When sufficient points of the workpiece feature surface are assessed, and the workpiece is correctly adjusted (e.g. with an adjustable support in order to avoid the effect shown in Fig. 18.70) this is an inspection close to the definition.

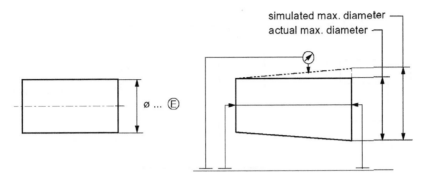

Fig. 18.70 Inspection of the envelope requirement: failure caused by insufficient adjustment of the workpiece feature

18.7.11.2.5 Inspection with length measuring instrument, measuring table and/or V-block

Revolutions of the workpiece feature in V-blocks or revolutions (rolling) on a measuring table cannot detect certain types of form deviations (oval and lobed), and should be applied only when these types of form deviations are not dominant (see 18.7.4.3).

18.7.11.2.6 Inspection with measuring table or measuring plate and length measuring instrument

The inspection method is shown in Figs 18.71 and 18.72, which show that it is necessary to inspect from both sides. With outer dimensions (Fig. 18.71), the method with

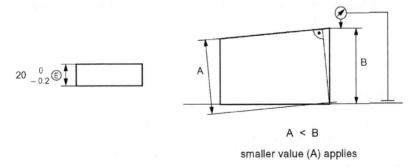

A < B

smaller value (A) applies

Fig. 18.71 Inspection of the envelope requirement of two parallel opposite plane surfaces on a measuring table with length measuring instrument: A < B; smaller value (A) applies

the smaller maximum indication applies and with inner dimension (Fig. 18.72), the method with the larger minimum indication applies.

When sufficient points of the workpiece feature surfaces are assessed, this is an inspection close to the definition.

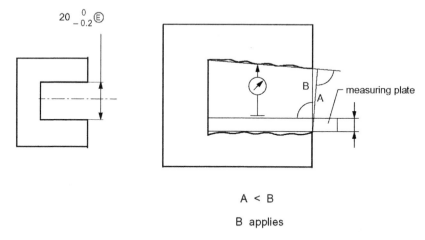

A < B

B applies

Fig. 18.72 Inspection of the envelope requirement of two parallel opposite plane surfaces with a measuring plate (e.g. gauge block) and a length measuring instrument: $A < B$; larger value (B) applies

18.7.12 Inspection of the maximum material requirement

18.7.12.1 Definition

It is to be checked whether the surface of the workpiece feature (cylindrical surface or two parallel opposite plane surfaces) is contained in the (almost) geometrical ideal boundary of maximum material virtual size (embodiment, reference element).

The reference elements at the toleranced feature and at the datum feature have the (almost) geometrical ideal orientation (parallel, perpendicular, in specified angle) and, when the maximum material requirement is applied to a location tolerance, are located (almost) at the theoretical exact (geometrical ideal) location.

The reference element at the toleranced feature is to be adjusted according to the datum or datum system (parallel, perpendicular, in specified angle). The orientation of the datum is defined by the minimum rock requirement at the datum feature of the workpiece (see 18.3.2) or by the adjustment of the workpiece in the datum system (see 3.4). The distance between reference element and datum (theoretical exact location of the reference element) is zero (coaxiality, symmetry) or specified by a theoretical exact (rectangular framed) dimension. The same applies to the distances between the reference elements.

When the maximum material requirement is applied to the datum feature, the reference element at the datum feature has (almost) geometrical ideal form at maximum material virtual size (see 9.2 and 9.3.3) and replaces the datum.

18.7.12.2 Inspection methods

The reference element (embodiment of the boundary of maximum material virtual size) may be established by

- functional gauge;
- calculation of the rectangular coordinates related to the straight and perpendicular guides of the coordinate measuring machine;
- calculation of the cylinder coordinates related to the straight, perpendicular and circular guides of the form measuring instrument (occurs with coaxiality tolerances to which the maximum material requirement is applied).

As an approximation, the actual sizes of the features and the actual distances of the features can be assessed for the inspection.

18.7.12.2.1 Inspection with functional gauge

For sizes and forms of the gauges see 9.2 and 9.3.

When the gauge covers the entire surfaces of the workpiece features, this is an inspection close to the definition (Fig. 18.73).

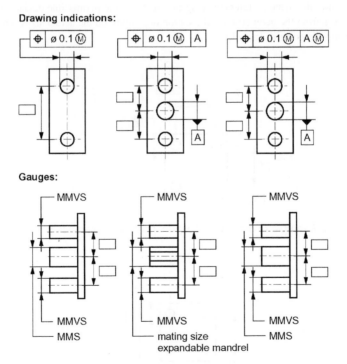

Fig. 18.73 Inspection of the maximum material requirement with functional gauges

18.7.12.2.2 Inspection with form measuring instrument or with coordinate measuring machine

When computer programs are applied to simulate the functional gauge, and sufficient sections or points of the feature's surfaces are assessed (probed), this is an inspection close to the definition.

18.7.12.2.3 Approximate inspection by measuring sizes and distances

When the deviations of form and orientation of the features are negligible compared with the location tolerances, the maximum material requirement applied to the location tolerance can be inspected by measuring the actual sizes of the toleranced feature(s), the actual sizes of the datum feature(s) and the actual distance(s) of the features.

The location tolerance may be exceeded by the difference between maximum material size and actual size (when the actual size does not take full advantage of the size tolerance).

18.7.12.2.3.1 Positional tolerances without datum. Figure 18.74 shows an example of four holes related to each other by positional tolerances but without datum. Figure 18.75 shows the relevant functional gauge (embodiments). Figure 18.76 shows the tolerance zones for the case when all actual sizes are at the maximum material size (ø 8.1) (and the holes almost of geometric ideal form and orientation related to each other). Figure 18.77 shows the dynamic tolerance diagram. It shows the permissible deviation of the actual axes from the geometric ideal location depending on the actual sizes. The same information is given in Table 18.5.

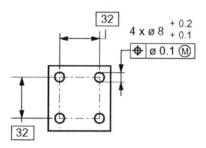

Fig. 18.74 Example of drawing indication, positional tolerancing without datum

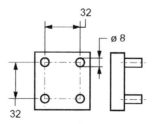

Fig. 18.75 Functional gauge according to the example in Fig. 18.74

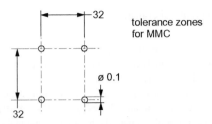

Fig. 18.76 Tolerance zones for the maximum material condition according to the example in Fig. 18.74

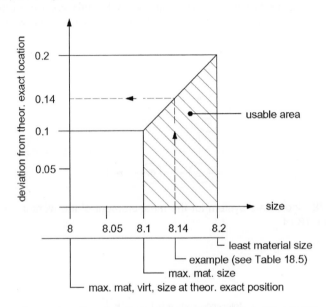

Fig. 18.77 Dynamic tolerance diagram for the example in Fig. 18.74

Table 18.5 Positional tolerances for the example in Fig. 18.74

Diameter	Positional tolerance
8.1	0.1
8.12	0.12
8.14	0.14
8.16	0.16
8.18	0.18
8.2	0.2

Measurements and graphical evaluations are as follows:

- specify the coordinate system for the measurement, e.g. the actual axis (centre) of the hole left below as coordinate origin, and the actual axis (centre) of the hole right below to determine the x axis (Fig. 18.78);
- assess the positional deviations of the holes in this coordinate system;
- plot the positional deviations on an enlarged scale (Fig. 18.79);
- use a transparent sheet (template) with the tolerance zone in the same enlarged scale and move it around until, if possible, all points are enclosed (Fig. 18.79);
- if a point remains outside the tolerance zone, measure the actual size of the feature (hole), enlarge the tolerance zone concentrically by the difference between maximum material size and actual size, and check whether the point is contained within this zone while the other points are contained in the (concentric) original tolerance zone (Fig. 18.79).

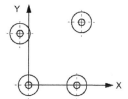

Fig. 18.78 Coordinate system for the measurement of the workpiece according to Fig. 18.74

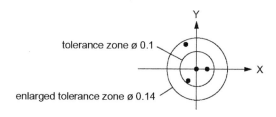

Fig. 18.79 Measured points (positional deviations) and tolerance template of the workpiece according to Fig. 18.74

18.7.12.2.3.2 Positional tolerances with datum. When the maximum material requirement applies to the datum, the pattern of tolerance zones (as a whole) may deviate from the theoretical exact location (relative to the datum axis) by the difference between the maximum material size and the actual size of the datum feature (this does not influence the location of the tolerance zones relative to each other).

Figure 18.80 shows an example of four holes related to each other by positional tolerances and related to a datum hole. The maximum material requirement applies to all five

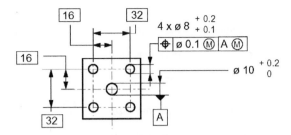

Fig. 18.80 Example of drawing indications, positional tolerancing with datum

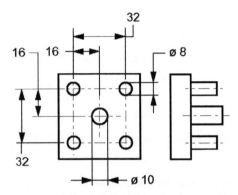

Fig. 18.81 Functional gauge according to the example in Fig. 18.80

holes. Figure 18.81 shows the relevant functional gauge (embodiment). Figure 18.82 shows the location and magnitude of the tolerance zones when

- the toleranced holes are at maximum material size, are at least material size;
- the datum hole is at maximum material size, is at least material size.

Table 18.6 gives the positional tolerances of the toleranced features (four holes) depending on their actual sizes, and the floating zone of the datum feature, depending on its actual size.

Measurements and graphical evaluations are as follows:

- specify the coordinate system for the measurement, e.g. the actual axis (centre) of the datum feature as coordinate origin, and the connection of the actual axes (centres) of the two holes below as the direction of the x axis (Fig. 18.83);
- assess the positional deviations of the holes in this coordinate system;
- plot the positional deviations in an enlarged scale (Fig. 18.84);
- use a transparent sheet (template) in the same scale with concentric
  - tolerance zone,
  - floating zone (difference between maximum material size and actual size of the datum feature)

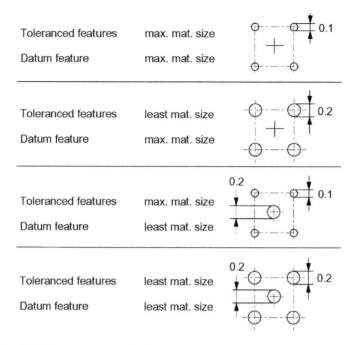

Fig. 18.82 Tolerance zones for the maximum material condition and for the least material condition of the example in Fig. 18.80

Table 18.6 Positional tolerances* for the example in Fig. 18.80

Diameter of toleranced hole	Positional tolerance of all holes	Diameter of datum hole	Floating zone of datum hole
8.1 MMS	0.1	10 MMS	0
8.12	0.12	10.05	0.05
8.14	0.14	10.1	0.1
8.16	0.16	10.15	0.15
8.18	0.18	10.2 LMS	0.2
8.2 LMS	0.2		

* Each combination of the values in the second and fourth columns is possible. The values of the second and fourth columns cannot be added, because they have different effects. Some extreme combinations are shown in Fig. 18.82.

and move it around until, if possible, all points are enclosed in the tolerance zone and the coordinate origin is still within the floating zone (Fig. 18.84);

- if a point remains outside the tolerance zone, measure the actual size of the feature (hole), enlarge the tolerance zone concentrically by the difference between maximum material size and actual size, check whether the point is contained within this

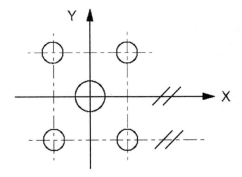

Fig. 18.83 Coordinate system for the measurement of a workpiece according to Fig. 18.80

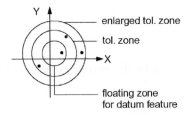

Fig. 18.84 Positional deviations and tolerance template for a workpiece according to Fig. 18.80

zone while the other points are contained in the (concentric) original tolerance zone and the coordinate origin in the floating zone (Fig. 18.84).

18.7.12.2.3.3 Positional tolerances with and without datum

Figure 18.85 shows an example of four holes related to each other by rather small positional tolerances and related to the datum features A and B (datum system) by rather large positional tolerances.

Figure 18.86 shows the relevant functional gauge (embodiment).

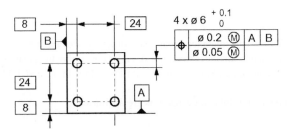

Fig. 18.85 Example of drawing indication, positional tolerances with and without datum

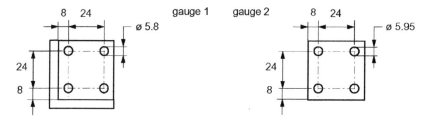

Fig. 18.86 Functional gauge according to the example in Fig. 18.85

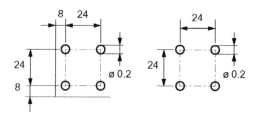

Fig. 18.87 Tolerance zones for the maximum material condition of the example in Fig. 18.85

tol. zones and actual positions

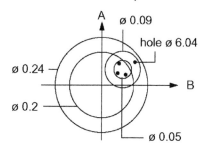

Fig. 18.88 Positional deviations and tolerance templates for a workpiece according to Fig. 18.85

Figure 18.87 shows the tolerance zones when all actual sizes are at the maximum material size (ø 6). Dynamic tolerance diagrams or tables containing the positional tolerances depending on the actual sizes of the holes similar to Fig. 18.77 and Table 18.5 could also be shown.

Measurements and graphical evaluations are as follows:

- assess the positional deviations of the holes in the coordinate system AB;
- plot the positional deviations in an enlarged scale (Fig. 18.88);

- use a transparent sheet (template) in the same scale with the tolerance zone ø 0.05 (Fig. 18.88), the procedure is the same as in 18.7.10.2.3.1;
- use a transparent sheet (template) in the same scale with the tolerance zone ø 0.2 with centre located at the coordinate origin, check whether all points are contained within this zone; if not, measure the actual size of the concerned hole, enlarge the tolerance zone by the difference between the maximum material size and the actual size, and check whether the point is within this zone (Fig. 18.88).

18.7.12.2.3.4 Positional tolerances for keyways. When gauges or coordinate measuring machines are not available the part may be inspected similar to Fig. 18.89 by simulating the gauge using a measuring plate, a measuring solid right angle and a dial gauge (see also Fig. 20.115).

18.7.12.2.4 Simplified inspection

When no functional gauges are available and coordinate measuring machines or form measuring instruments are not currently used and the prerequisites according to 18.7.10.2.3 do not apply, the workpieces may be inspected in the first step as if the maximum material requirement were not applied. Only when with this inspection the location tolerance is exceeded should it be checked in the second step by a suitable method (e.g. by a coordinate measuring machine), whether the maximum material requirement is violated.

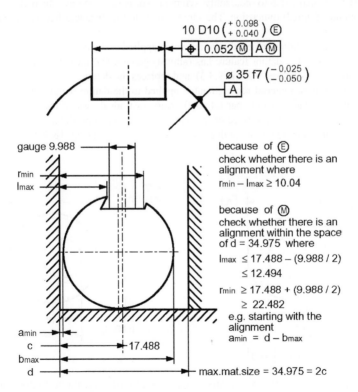

Figure 18.89 Inspecting a keyway by simulating the gauge

A prerequisite for this procedure is that the total tolerance is split into a size tolerance and a location tolerance (and is not indicated as size tolerance together with a zero location tolerance).

18.7.13 Inspection of the least material requirement

18.7.13.1 Definition

It is to be checked whether the surface of the workpiece feature (cylindrical surface or two parallel opposite plane surfaces) violates the (almost) geometrical ideal boundary of least material virtual size (embodiment, reference element).

The reference elements of the toleranced feature and of the datum feature have the (almost) geometrical ideal orientation (parallel, perpendicular, in specified angle) and, when the least material requirement is applied to a location tolerance, are located (almost) at the theoretical exact (geometrical ideal) location.

The reference element at the toleranced feature is to be adjusted according to the datum or datum system (parallel, perpendicular, in specified angle). The orientation of the datum is defined by the minimum rock requirement at the datum feature of the workpiece (see 18.3.2) or by the adjustment of the workpiece in the datum system (see 3.4). The distance between reference element and datum (theoretical exact location of the reference element) is zero (coaxiality, symmetry) or is specified by a theoretical exact (rectangular framed) dimension. The same applies to the distances between the reference elements.

When the maximum material requirement is applied to the datum feature, the reference element at the datum feature has (almost) geometrical ideal form at maximum material virtual size (see 9.2 and 9.3.3), and replaces the datum.

When the least material requirement is applied to the datum feature, the reference element at the datum feature has (almost) geometrical ideal form at least material virtual size (see 11.3). The surface of the datum feature must not violate this boundary. The minimum rock requirement does not apply to this datum (Fig. 18.90).

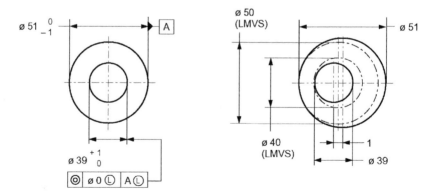

Fig. 18.90 Least material requirement applied to the toleranced feature and to the datum feature

18.7.13.2 Inspection methods

The reference element (embodiment of the boundary of least material virtual size) may be established by

- calculation of the rectangular coordinates related to the straight and perpendicular guides of the coordinate measuring machine;
- calculation of the cylinder coordinates related to the straight, perpendicular and circular guides of the form measuring instrument (occurs with coaxiality tolerances to which the least material requirement is applied).

As an approximation the actual sizes of the features and the actual distances of the features can be assessed for the inspection.

As it must be checked whether the reference element is entirely within the material of the workpiece feature (whether the workpiece feature's surface violates the reference element) gauging is not possible.

18.7.13.2.1 Inspection with coordinate measuring machine or with form measuring instrument

When computer programs are applied to simulate the geometrical ideal boundary of least material virtual size, and sufficient points or sections of the feature's surfaces are assessed, this is an inspection close to the definition.

18.7.13.2.2 Approximate inspection by measuring sizes and distances

When the deviations of form and orientation of the features are negligible compared with the location tolerances, the least material requirement applied to the location tolerance can be inspected by measuring the actual sizes of the toleranced feature(s), the actual sizes of the datum feature(s) and the actual distances of the features.

The location tolerance may be exceeded by the difference between the least material size and the actual size (when the actual size does not take full advantage of the size tolerance).

Figure 18.91 shows an example of application of the least material requirement. Figure 18.92 shows the relevant geometrical ideal boundary at the least material virtual size that is coaxial to the datum A (reference element) (see 11.3).

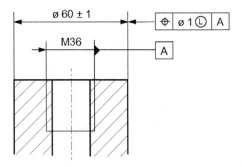

Fig. 18.91 Example of drawing indication, positional tolerancing applied to the least material requirement

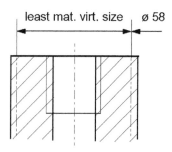

Fig. 18.92 Geometrical ideal boundary at least material virtual size for the example in Fig. 18.91

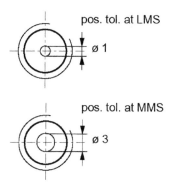

Fig. 18.93 Positional tolerance zones at maximum material condition and at least material condition according to the example in Fig. 18.91

Table 18.7 Positional tolerances for the example in Fig. 18.91

Diameter of cylinder	Positional tolerance of cylinder
59 LMS	1
60	2
61 MMS	3

Figure 18.93 shows the positional tolerance zone at least material size and at maximum material size. Table 18.7 lists the diameters of the positional tolerance zones of the outer cylindrical feature depending on its actual sizes.

The positional deviation may be assessed as described in 18.7.9 (see, e.g., Fig. 18.56).

18.7.14 Assessment of the positional deviation for projected tolerance zones

18.7.14.1 Definition

The deviations of a specified extension (projection) of the feature axis from an (almost) geometrical ideal reference straight line (embodiment, simulated straight line, ISO 5459), which is in the (almost) geometrical ideal orientation and location, is measured.

The positional deviation δ_c is the largest distance (within the specified length) of the extension of the axis from the reference straight line, and must not exceed half the positional tolerance t_c: $\delta_c \leq t_c/2$.

18.7.14.2 Measuring methods

For the measurement, the extension of the axis to be assessed can be established, for example by

- fitting, as far as possible without clearance, with an (almost) geometrical ideal counterpart (conical threaded mandrel, screw, bolt);
- simulation of the fitted counterpart (by calculation) in the coordinate measuring machine.

The positional deviation along the specified extension of the feature axis is measured as described in 18.7.9.3. See also 7.

18.8 Assessment of geometrical deviations of threaded features

If not otherwise specified, the tolerances of orientation, location or run-out of threaded features apply to the pitch diameter (ISO 1101).

For the measurements are used:

- form measuring instruments*;
- coordinate measuring machines*;
- special devices (see, e.g., Figs 18.94 to 18.97).

When measuring deviations of orientation, location or run-out, special measuring devices are to be applied for the support in the thread with or without clearance (with or without application of the maximum material requirement) and for probing or tracing of threaded features.

For the support in the thread without clearance (datum without application of the maximum material requirement), the following may be used:

- conical threaded mandrel (e.g. with a cone angle of 0.5°);
- mandrel or ring with two parts of thread that can be adjusted in radial or axial direction in order to contact the thread flanks without clearance;
- two thread ring gauges screwed against each other

with surfaces to be supported and sufficiently cylindrical and coaxial with the thread.

The same devices may be used at the toleranced threaded features to assess the deviations of orientation or location or run-out by measurement at the cylindrical surfaces (see 18.7.8, 18.7.9 and 18.7.10).

Instead of expandable or conical or parted thread devices (mandrels) the revised standard ISO 4759-1 on tolerances for fasteners will probably recommend the use of (lubricated) spring washers to support the proper alignment of go gauge thread devices along the workpiece thread flanks (see Figs 18.94, 18.95, 18.97 and 18.99).

---

* If suitable probes and computer programs are available or if measured together with special devices as described in this section.

For probing at the pitch diameter by a self-centring probing mode, the following may by used:

- ball of diameter suitable for the thread pitch (Fig. 18.98 and Table 18.8);
- notch and/or measuring cone suitable for the thread pitch (median diameter of the truncated cone = *P*/2) (see Fig. 18.98);
- thread segments.

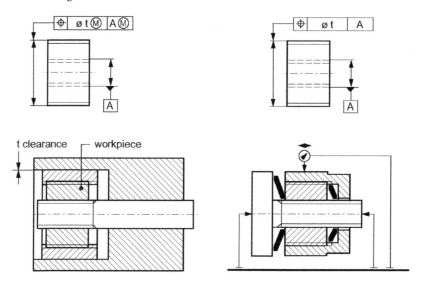

Fig. 18.94 Assessment of the coaxiality deviation of a threaded ring

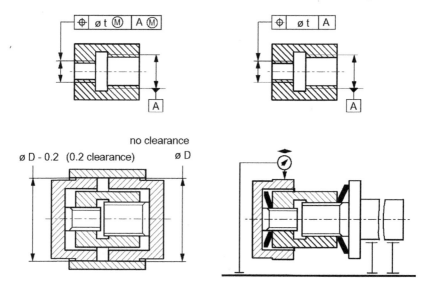

Fig. 18.95 Assessment of the coaxiality deviation of a threaded ring

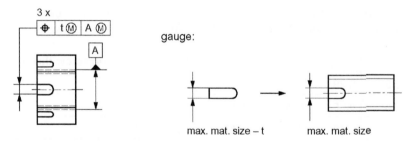

Fig. 18.96 Gauge for the inspection of the symmetry deviation of the slot relative to the thread of a nut

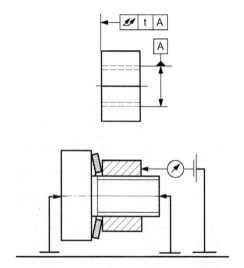

Fig. 18.97 Assessment of the axial run-out deviation of a nut

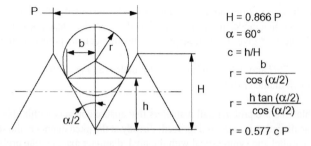

$H = 0.866\ P$

$\alpha = 60°$

$c = h/H$

$r = \dfrac{b}{\cos(\alpha/2)}$

$r = \dfrac{h\tan(\alpha/2)}{\cos(\alpha/2)}$

$r = 0.577\ c\ P$

Fig. 18.98 Metrical thread, configuration, probing ball radius *r* for probing near the flanks (using a self-centring probing mode)

Table 18.8 Recommended probing ball diameters D_{rec} in mm for probing at the pitch diameters of metric threads according to ISO 261 (P pitch in mm)

P	$D_{0.5H}$	$D_{0.375H}$	$D_{0.75H}$	D_{rec}	ISO metric screw threads
0.5	0.29	0.22	0.43	0.3	M3
0.6	0.35	0.26	0.52	0.3	M3.5
0.7	0.40	0.30	0.61	0.5	M4
0.75	0.43	0.32	0.65	0.5	M4.5
0.8	0.46	0.35	0.69	0.5	M5
1	0.58	0.43	0.87	0.5	M6 M7
1.25	0.72	0.54	1.08	0.8	M8 M9
1.5	0.87	0.65	1.30	0.8	M10 M11
1.75	1.01	0.76	1.51	1	M12
2	1.15	0.87	1.73	1	M14 M16
2.5	1.44	1.08	2.16	1.5	M18 M20 M22
3	1.73	1.30	2.60	1.5	M24 M27
3.5	2.02	1.52	3.03	2	M30 M33
4	2.31	1.73	3.46	2	M36 M39
4.5	2.60	1.95	3.89	3	M42 M45
5	2.89	2.17	4.33	3	M48 M52
5.5	3.18	2.38	4.76	3	M56 M60
6	3.46	2.60	5.19	3	M64 to M120
8	4.62	3.46	6.92	5	M125 to M180

When deviations of orientation or location are to be measured with a probing ball by a self-centring probing mode, the probing ball should contact the thread flanks near the pitch diameter. Table 18.8 gives the ball diameters when the contact points are theoretically

- between $0.75H$ and $0.375H$ of the thread profile (see Fig. 18.98);
- at the pitch diameter ($0.5H$).

D_{rec} in Table 18.8 are probing ball diameters that ensure contacts within this range.

When the deviation is measured (probed) directly at the pitch diameter, the stylus must be guided parallel and symmetrical with the pitch diameter axis, i.e. the probing direction line (axis) must meet the pitch diameter axis. With coordinate measuring machines, this guidance is not necessary when it is replaced by appropriate calculations.

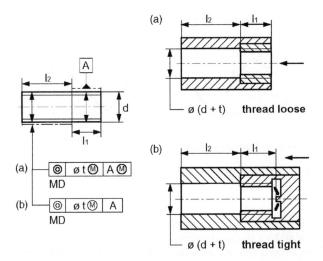

Fig. 18.99 Threaded part with maximum material requirement applied to achieve suitability for a stud function (projected tolerance zone for the counterpart, Fig. 7.7) drawing indications and gauges

When coordinate measuring machines and a small ball probe (smaller than according to Table 18.20) are used the thread flanks are to be assessed in order to calculate the substitute thread flanks (see 8.1). The line of intersection of two adjacent substitute thread flanks is used in order to calculate the pitch diameter line according to the theoretical thread configuration (Fig. 18.98).

For the inspection of threaded features to which the maximum material requirement is applied, gauges with threads (almost) at the maximum material size are to be applied (see 9 and Fig. 18.99), or when coordinate measuring machines are applied and appropriate software is available, the gauges may be simulated.

18.9 Tracing and probing strategies

18.9.1 General

The workpiece features can be inspected by

- gauging, e.g. with a functional gauge (covering the entire feature);
- areal contacting, e.g. with a measuring plate or mandrel;
- tracing, scanning (continuously, consecutive), e.g. with a dial gauge or form measuring instrument;
- probing (assessment of a number of points) (discontinuously), e.g. with a coordinate measuring machine.

With tracing, scanning and probing, the surface is assessed by sampling at selected sections or points. Therefore these methods are approximate. The more sections or points are assessed, the more realistically are the geometrical deviations assessed and the less is the measurement uncertainty.

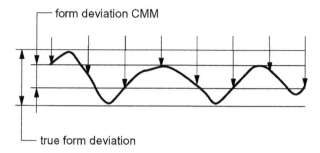

Fig. 18.100 Incomplete assessment of form deviations because of too wide spacing of assessed points

Figure 18.100 shows form deviations whose amplitude cannot be fully assessed because the spacing of probing (assessed points) is too wide.

In order to assess a certain bandwidth of deviations the Nyquist theorem should be respected in all directions of the surface, i.e. the spacing of the assessed points including the distances of sections to be traced or scanned should be at least not more than one half (better not more than one seventh) of the deviation wavelength to be assessed.

Figure 18.101 shows the effect of the spacing c of the assessed points on a surface exhibiting sinusoidal form deviations of the wavelength λ. When the spacing c is greater than half of the wavelength λ the wavelength of the sine wave cannot be assessed (Fig. 18.101 (a)). When the spacing is 0.4λ the wavelength of the sine wave will be assessed. But some amplitudes will be considerably reduced (Fig. 18.101(b)). When the spacing c is 0.2λ the true shape of the sine wave (wavelength and amplitudes) is almost obtained (Fig. 18.101(c)).

With present techniques, this would lead to very time-consuming and costly inspections. Therefore, for economical reasons, the number of selected sections and the number of probed points are normally reduced, but should be distributed in an optimized way. The necessary numbers of points and sections to be assessed and their optimized distribution depend on:

- the type (shape) of the form deviation (depending on the type of manufacturing process);
- the magnitude of the form deviation;
- the ratio of form deviation to geometrical tolerance;
- the geometrical characteristic to be assessed.

Normally for the assessments of, for example, a datum axis or circular run-out deviations, fewer sections are necessary than for the assessments of cylindricity or total run-out deviations.

International Standards on this subject do not yet exist. However, the British Standard BS 7172 and former East German standards TGL 39 093 to TGL 39 098 and TGL 43 041 to 43 045 give some recommendations, which are included in the following. They may serve as a guide in cases where there is no specific information available on the features to be measured that would lead to different strategies (e.g. information on the manufacturing process and on the type and magnitude of the form deviations).

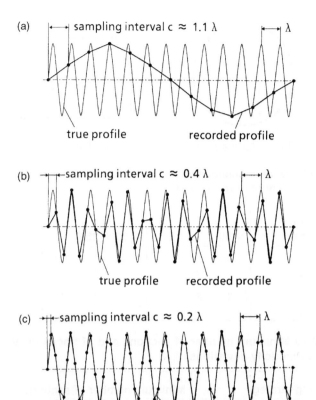

Fig. 18.101 Assessment of sinusoidal form deviations depending on the spacing c of assessed points relative to the wavelength λ of the sine wave

18.9.2 Tracing strategies

Flatness: Figure 18.102 shows recommended traces for the assessment of flatness deviations.

Roundness: Table 18.9 gives the recommended number and location of radial sections (measuring sections) for the assessment of roundness deviations, according to TGL 39 096.

The distance between the measuring sections is l/N, and the distance between the first or last measuring section and the end of the feature is $l/2N$.

Cylindricity: Table 18.10 gives the recommended number of sections and number of traces for the assessment of cylindricity deviations, according to TGL 39 097.

Datum axis: Table 18.11 gives the recommended number and location of radial sections (measuring sections) for the assessment of datum axes, according to TGL 43 043.

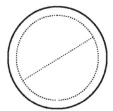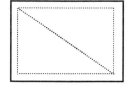

Fig. 18.102 Traces for the assessment of flatness deviations

Table 18.9 Recommended number and location of measuring sections for the assessment of roundness

Length *l* of surface mm	≤ 50			> 50 ≤ 250			> 250		
Ratio *l/d* of length/diameter	≤ 1	> 1 ≤ 3	> 3	≤ 1	> 1 ≤ 3	> 3	≤ 1	> 1 ≤ 3	> 3
Number *N* of sections	1	2	3	2	3	4	3	4	5

Table 18.10 Minimum number of sections and of traces for the assessment of cylindricity deviations

Measuring strategy	Minimum number of sections	Lines
Radial section method	3 radial sections	3
Generatrix method	3 axial sections	6
Helical method	2 radial sections + 1 helical line of 2 pitches	3
Extreme positions method	1 axial section + 2 radial sections	4

18.9.3 Probing strategies

The points should be distributed over the entire surface, but not so that they could follow systematic periodic form deviations. For example, six points equally spaced on the circumference of a cylinder cannot detect three lobed form deviations (Fig. 18.103). However, seven points equally spaced on the circumference of a cylinder can assess these form deviations by 79% of their amplitudes.

Straight line: The length is to be divided into $3N - 2$ subintervals, with points placed in the 1st, 4th, 7th, ..., $(3N - 2)$ subintervals, at a random positions (Fig. 18.104), where N is the number of points.

Plane: The area is to be divided into $N_1 \times N_2$ rectangles (squares, if possible). Within each rectangle, one point is placed at a random position (Fig. 18.105).

Table 18.11 Recommended number and location of the measuring sections for the assessment of datum axes

Length L_B of datum cylinder (mm)	$\leqslant 50$	$> 50 \leqslant 250$	> 250
Ratio L_B/d_B of datum cylinder	$\leqslant 3 > 3$	$\leqslant 1 > 1 > 3$ $\leqslant 3$	$\leqslant 1 > 1 > 3$ $\leqslant 3$
Number of radial sections n_B	2 3	2 3 4	3 4 5
Face distance of radial sections (mm)	$L_B/8$ $L_B/12$	$L_B/8$ $L_B/12$ $L_B/16$	$L_B/12$ $L_B/16$ $L_B/20$
Distance between radial sections (mm)	$3L_B/4$ $5L_B/12$	$3L_B/4$ $5L_B/$ 12 $7L_B/24$	$5L_B/12$ $7L_B/$ 24 $9L_B/40$

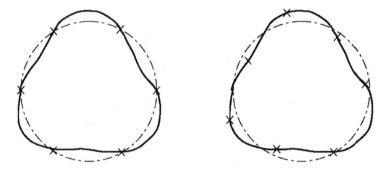

Fig. 18.103 Distribution of points with lobed form deviations

Fig. 18.104 Distribution of $N = 5$ points on a straight line

When only a small number of points are to be measured, these points may be placed in alternate rectangles in a "chessboard" fashion (Fig. 18.106).

Circle: N equally spaced points are used, where N is, if possible, a prime number and greater than the expected number of lobes (if a lobed form is to be expected) (Fig. 18.103).

Sphere: For a sphere sector between two parallel planes, N points are distributed over n_c sections parallel to the end faces and equally spaced with n_p points each (Fig. 18.107),

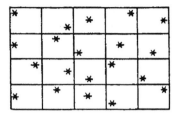

Fig. 18.105 Distribution of $N = 20$ points on a plane surface

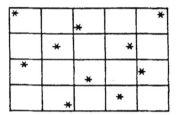

Fig. 18.106 "Chessboard" fashion distribution of $N = 10$ points on a plane surface

Fig. 18.107 Distribution of $N = 30$ points on a spherical surface of $r = 100$ mm and $h = 150$ mm, so that $n_c = 3$ and $n_p = 10$

where
h = height of sphere section in mm
r = sphere radius in mm
n_c = number of sections parallel to the end faces (including the latter)
n_p = number of points in each section
N = total number of points

$n_c \approx \sqrt{[Nh/(2\pi r)]}$
$n_p \approx N/n_c$

For a complete sphere single points at each pole are also to be measured.

Cylinder:
(a) Dividing similarly to the plane;
(b) Dividing similarly to the sphere.

It is recommended to alternate between odd and even numbers n_p of points on the circles, in order to detect lobings (Fig. 18.108).

Fig. 18.108 Distribution of 30 points on a cylindrical surface of $r = 10\,mm$ and $h = 30\,mm$, ($n_c = 4$, $n_p = 7$ or 8)

Fig. 18.109 Distribution of $N = 35$ points over the surface of a truncated cone of $r_1 = 10\,mm$, $r_2 = 15\,mm$, $h = 20\,mm$ ($k = 20.6\,mm$), so that $n_c = 3$, $s = 2$, $n_p = 10$, 12 and 14

Cone: N points are used, distributed over n_c sections perpendicular to the axis and equally spaced, where

h = height of truncated cone in mm
r_1 = radius at the smaller end in mm
r_2 = radius at the larger end in mm
k = length of circumference in mm
n_c = number of sections (end faces included)
n_p = number of points of a section (this should decrease towards the vertex of the cone
    by the number s; Fig. 18.109).
k = $\sqrt{[h^2 + (r_2 - r_1)^2]}$

$n_c \approx \sqrt{[kN/\pi(r_1 + r_2)]}$
$s \approx 2\pi(r_2 - r_1)/k$

18.9.4 Number of points

In the absence of form deviations, a minimum number of points would be sufficient to determine the geometrical feature in a coordinate system (Table 18.12). Because there are always form deviations, more points have to be assessed. Table 18.12 gives the minimum numbers recommended by BS 7172. Fewer points should not be chosen. More points decrease the measurement uncertainty caused by the form deviations.

Table 18.12 Recommended number of points

Feature	Number of probed points		Remarks
	Mathematical	Recommended	
Straight line	2	5	
Plane	3	9	Distributed on three lines
Circle	3	7	For assessment of three-lobed forms
Sphere	4	9	Distributed on three parallel sections
Cylinder	5	12	For assessment of straightness distributed on four radial sections
		15	For assessment of roundness distributed on three radial sections
Cone	6	12	For assessment of straightness distributed on four radial sections
		15	For assessment of roundness distributed on three radial sections

In the literature (see Ref. [8]) there is another recommendation to choose at least $8n$ points, where n is the mathematical minimum number of points.

For curved features with changing curvature (e.g. turbine blades) the spacing of the points should be closer for regions of small radii of curvature than for regions of larger radii of curvature.

18.10   Separation from roughness and waviness

The separation of roughness and waviness from geometrical deviations (form, orientation, location, run-out) is not yet completely internationally standardized.

Normally the peaks of the surface roughness contribute fully to the geometrical deviation. How much the surface roughness valleys contribute to the geometrical deviation depends on the measuring method.

In practice the roughness is filtered out by the effect of the stylus tip (ball) of the measuring instrument (with dial gauges, the ball radius is normally 1.5 mm). The effect on the assessment of the geometrical deviation depending on the ball radius r and on the spacing c of the deviations (irregularities) of the workpiece surface is shown in Fig. 18.110. With surfaces manufactured by metal removal normally the waviness depth W_t is

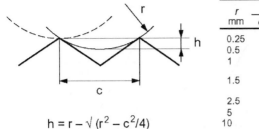

r	h in μm with c in mm			
mm	$c = 0.01$	$c = 0.1$	$c = 0.3$	$c = 1$
0.25	0.05	5	50	
0.5	0.025	2.5	23	
1	0.0125	1.25	11	134
1.5	0.0083	0.83	7.5	86
2.5	0.0050	0.50	4.5	51
5	0.0025	0.25	2.3	25
10	0.0013	0.13	1.1	13

$$h = r - \sqrt{(r^2 - c^2/4)}$$

Fig. 18.110 Effect of the stylus ball radius r and of the workpiece deviation spacing c on the assessment of geometrical deviations

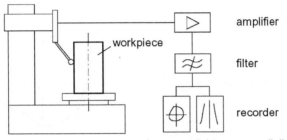

roundness, straightness, parallelism

Fig. 18.111 Form measuring instrument

smaller than 5 μm and the waviness spacing wider than 0.3 mm. In these cases the waviness is part of the geometrical deviations. Therefore the corresponding ISO Technical Committee plans to define in an ISO Standard that waviness is part of the geometrical deviations.

However, for some functions, e.g. the ball race surface on ball bearings, the waviness must be toleranced and measured separately by the use of appropriate filters, see Fig. 18.115.

Form measuring instruments usually allow the use a low-pass filter to separate (filter out) roughness (and parts of waviness) in the assessment of geometrical deviations. Figure 18.111 shows the composition of the instruments. Figure 18.112 shows the filter characteristics of these instruments.

The roundness profile of a workpiece can be considered as a superposition of sine waves. Fig. 18.113 shows such a Fourier analysis of a roundness profile.

Each sine wave portion will be transmitted by the filter according to the filter characteristic with full or attenuated amplitudes. Waves of short wavelengths (several waves per revolution) (e.g. roughness) will not be transmitted by the filters (or will only be transmitted with strong attenuation).

The filters are named according to the limiting number of waves (cut-off) n_g or according to the limiting wavelength (cut-off) λ_R. These are the number of sine waves (undulations) per revolution (circumference) (UPR) or the sine wave length where the amplitudes are transmitted by 75% (old RC filter according to ISO 3274–1975) or by 50% (new

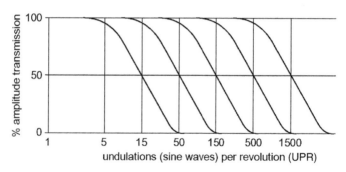

Fig. 18.112 Filter characteristics

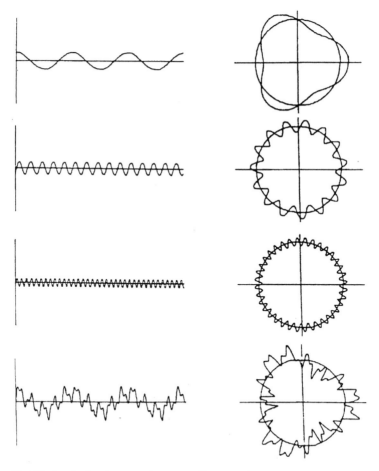

Fig. 18.113 Analysis of a roundness profile according to Wirtz (see Ref. [11])

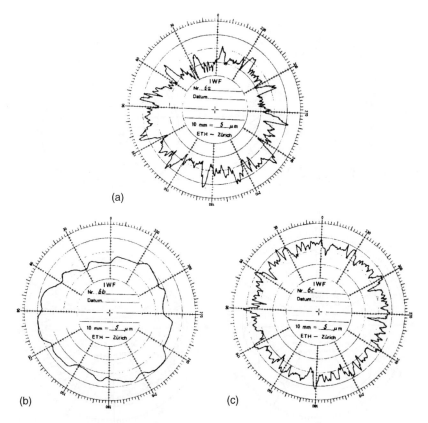

(a)

(b)   (c)

Fig. 18.114 Effect of different filters: (a) without filter; (b) with filter 50; (c) with a filter combination (band pass) (according to Wirtz [11])

phase correct Gauss filter according to ISO 11 562). The measuring results may be significantly different depending on whether the RC filter or the Gauss filter has been used.

Figure 18.114 shows an example of the effect of the different filters on the same workpiece profile to be measured. It shows that the result of the measurement is also significantly different depending on the filter size (cut-off) that has been used.

Using narrow band-pass filters (combination of low-pass and high-pass filters) deviations of certain wavelengths can be selected. Thereby eventually the reasons for certain form deviations and the reasons for the functional behaviour become evident. Figure 18.115 shows such an analysis. The three-lobed form indicates the effect of the three-point chuck.

In cases of dispute as to whether a workpiece does or does not comply with the geometrical tolerance, the rate of included waviness may be decisive. In these cases it is recommended to agree on the following:

- when measuring instruments with filter devices are available, to apply a stylus tip ball radius of 0.5 mm and a Gauss filter of 0.8 mm for straight features and according to Table 18.13 for round features;

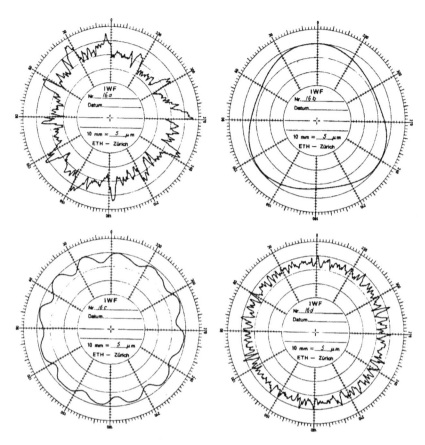

Fig. 18.115 Measurement of the same profile with different band passes (according to Wirtz [11])

Table 18.13 Filter (ISO 11 562) in UPR for the measurement of roundness and cylindricity according to ISO 12 181* and ISO 12 180*

Workpiece diameter (mm)			Filter UPR
	0 to	8	15
above	8 to	25	50
above	25 to	80	150
above	80 to	250	500
above	250		1500

* In preparation.

Table 18.14 Low-pass cut-off λ_R for the assessment of straightness and flatness deviations according to TGL 43 041

Roughness R_a (μm)	Roughness R_z (μm)	Cut-off λ_R (mm)
$\leqslant 0.025$	$\leqslant 0.1$	0.25
$>0.025 \leqslant 0.4$	$>0.1 \leqslant 1.6$	0.8
$>0.4 \leqslant 3.2$	$>1.6 \leqslant 12.5$	2.5
$>3.2 \leqslant 12.5$	$>12.5 \leqslant 50$	8
$>12.5 \leqslant 100$	$>50 \leqslant 400$	25

(Then the height of narrow peaks may be considerably attenuated by the Gauss filter. In order to limit the height of these peaks on the workpiece an additional specification on waviness may be necessary. Application of stylus tips without further filtering (see below) should assess the full peak height and may be preferable.)
- when only measuring instruments without filter devices are available or shall be applied, a stylus tip ball radius of 1.5 mm. (For small holes and slots a smaller stylus tip ball radius must be agreed upon.)

This is what will probably be standardized by ISO. If, for example, in international trade other specifications shall be applied (e.g. according to national standards), this should be specified in the drawing or in related documents.

The filters according to Table 18.13 have been calculated so that for a median diameter a filter has a similar effect as a 0.8 mm filter for straight features. The 0.8 mm filter has been chosen because it is the filter that is most frequently used for roughness measurements according to ISO 4288 and for waviness measurements. The tip radius 0.5 mm has been chosen in order to leave its filter effect beyond the 0.8 mm filter effects, i.e. what is filtered out by the tip radius will be filtered out by the 0.8 mm filter anyway, so that the tip radius has no filtering effect on the measurement result.

ANSI B89.3.1 states that, if not otherwise specified, for round features 0.25 mm tip radius and the 50 UPR filter apply (see 21.1.8).

In the past in the East European countries the filters recommended for measurement with form measuring instruments are given in Table 18.14 for the measurement of straightness deviations and in Table 18.15 for the measurement of roundness, cylindricity, coaxiality and circular radial run-out deviations according to the former East German Standards TGL 43 041, TGL 39 096, TGL 39 097, TGL 43 042 and TGL 43 043. Similar tables for the measurement of axial run-out deviations were given in TGL 43 044.

When the stylus tip (ball) penetrates into the grooves of the roughness (e.g. with turned surfaces), it may occur that the groove is inclined to the section plane of probing so that the stylus tip goes from the peak to the valley within, for example, half of a revolution. Thereby an eccentricity is simulated that does not in fact exist (Fig. 18.116). This eccentricity remains even when there is a filter involved that separates the roughness from the geometrical deviation. This error will be avoided by the use of a suitable stylus tip form (e.g. a hatched form with radius $r = 10$ mm; Figs 18.116 and 18.117).

Table 18.15 Low-pass cut-off n_g for the assessment of roundness, cylindricity, coaxiality, circular radial run-out and total radial run-out deviations according to TGL 39 096

Nominal diameter d (mm)	Cut-off n_g Undulations per revolution (UPR)			
	$\leqslant 2.5$	> 2.5 $\leqslant 5$	> 5 $\leqslant 10$	> 10
$\leqslant 10$	150	50	50	50
$> 10 \leqslant 50$	500	150	150	50
$> 50 \leqslant 120$	1500	500	500	150
$> 120 \leqslant 250$	1500	1500	500	500
> 250	1500	1500	1500	1500

When $n_g = 1500$ UPR is not achievable with the measuring instrument it is permissible to choose the greatest possible cut-off.

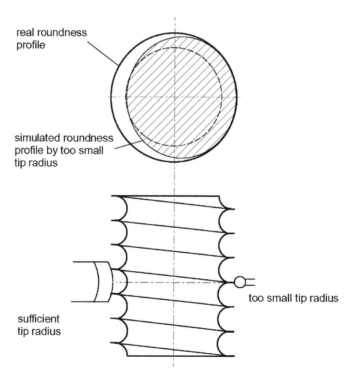

Fig. 18.116 Simulated eccentricity of a turned surface

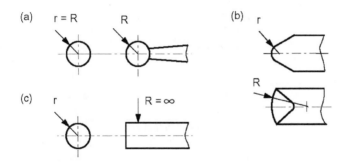

Fig. 18.117 Stylus tip forms: (a) ball; (b) hatched (toroidal); (c) cylindrical form

18.11 Measurement uncertainty

18.11.1 Definition

According to the International Vocabulary of Basic and General Terms in Metrology (VIM) the measurement uncertainty is an estimate characterizing the range of values within which the true value of a measurand lies.

The measurement uncertainty comprises many components, i.e. the random deviations (errors) and the unknown (and therefore uncorrected) systematic deviations (errors) of all quantities that contribute to the measuring result.

According to ISO 14 253-1 for geometrical deviations (Geometrical Product Specifications (GPS)) for measurement uncertainty the statistical confidence level of 95% applies, if not otherwise specified.

18.11.2 Application

ISO 9001:1994 clause 4.11.1 specifies:

Inspection, measuring and test equipment shall be used in a manner which ensures that the measurement uncertainty is known and is consistent with the required measuring capability.

The standard ISO 14 253-1 defines uncertainty zones at the specification (tolerance) limits. The width of the uncertainty range (zone) is ± the expanded measurement uncertainty U, see Fig. 18.118. For tolerances of form, orientation, coaxiality and symmetry, and run-out, with the one tolerance limit zero, there is only one uncertainty zone at the other tolerance limit, see Fig. 18.119.

The standard specifies the following decision rules for the case where the result of measurement falls within one of the uncertainty zones:

- Repeat the measurement with a measurement process with a smaller measurement uncertainty or repeat estimating the measurement uncertainty more precisely; if at the end the measurement result still falls within an uncertainty zone the following rule applies:
- the supplier cannot prove conformance with the specification (tolerance) in the case of an outgoing inspection; the customer cannot prove non-conformance with the

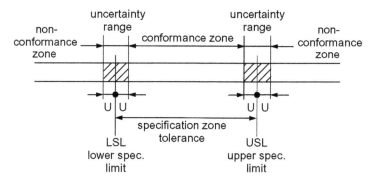

Fig. 18.118 Conformance zone, non-conformance zones and uncertainty zones (ranges) with specification zone limits, both ≠ 0

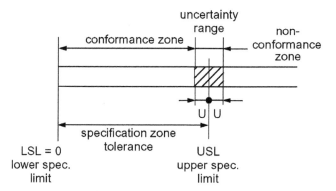

Fig. 18.119 Conformance zone, non-conformance zone and uncertainty zone (range) with one specification limit = 0

specification (tolerance) in the case of an incoming inspection. Resellers should use the proof provided to them by their supplier in order to avoid the situation where they cannot reject the delivery but they also cannot submit the delivery.

18.11.3 Assessment

The GUM Guide to the expression of uncertainty in measurement (European standard EN 13 005) applies.

The GUM has been elaborated by international organizations and has been adopted by most of the national laboratories for measurement and standards (e.g. NIST in the USA, NPL in the UK, PTB in Germany). GUM gives the rules as to how to assess the measurement uncertainty correctly.

It distinguishes between:

Evaluation type A: using statistical means (repeated measurements)
Evaluation type B: using uncertainty components (investigations)

The bases for evaluation type B are:

- data from former measurements;
- experimental results, investigations;
- indications of the measuring device manufacturer;
- dates of calibration.

There is no longer a distinction between random deviations and uncorrected systematical deviations.

GUM is (because of its generality) very theoretical, very voluminous and therefore not easy to read and difficult to understand and implement into industrial practice.

Therefore, for geometrical product specifications (GPS) (specifications for dimensions, geometry, roughness and waviness), the ISO Technical Report ISO TR 14 253-2 has been developed. This report contains a simplified procedure for the assessment of the measurement uncertainty and a simplified procedure for optimizing the cost of manufacturing and inspection (PUMA: procedure for uncertainty management).

The simplifications are in a conservative way so that they result in larger measurement uncertainty values (safe method).

The simplifications are:

- the sensitivity coefficients (see GUM) are equal to 1;
- the correlation coefficients ρ are either $\rho = 1, -1, 0$; if the uncertainty components are not known to be uncorrelated, full correlation is assumed either $\rho = 1$, or $\rho = -1$;
- restriction to three distribution types (normal-, rectangular-, U-distribution);
- conservative assumption of the distribution type; if not known to be normal take rectangular or U-distribution; if not known to be at least rectangular take U-distribution;
- start the iteration procedure with evaluation type B.

The procedure of uncertainty budgeting is based on the error propagation law.

For the **black box method** (direct measurement) the combined standard uncertainty (standard deviation) u_c is

$$u_c = \sqrt{(\Sigma\, u_{xr})^2 + (\Sigma\, u_{xp})^2}$$

where u_{xr} = standard deviation of correlated uncertainty contributor

u_{xp} = standard deviation of uncorrelated uncertainty contributor

For the **transparent box method** (indirect measurement) when the geometrical deviation is assessed by indirect measurements, i.e. the deviation δ to be measured is calculated from the measured quantities $(x_1, x_2, \dots x_i, \dots x_{p+r})$ according to the function

$$\delta = G(x_1, x_2, \dots x_i, \dots x_{p+r})$$

The combined standard uncertainty of measurement u_c is

$$u_c = \sqrt{u_r^2 + \sum_{i=1}^{p}\left(\frac{\partial\delta}{\partial x_i}\, u_{xi}\right)^2} \qquad U_r = \sum_{i=1}^{r}\frac{\partial\delta}{\partial x_i}\, u_{xi}$$

where u_r = the contribution ("sum") of the correlated components of the measurement uncertainty

u_{xi} = the combined uncertainty of measurement of the number i measured value (function) that is part of the transparent box method of uncertainty estimation for the measurement of δ; u_{xi} can be the result (u_c) of either a black box or another transparent box method of uncertainty estimation

r = the number of correlated components of the measurement uncertainty

p = the number of uncorrelated components of the measurement uncertainty

The uncertainty contributors u_x are due to the influences on the measuring process and are to be assessed as standard deviations in length. When only the limit values are known (range a) the standard deviation of the uncertainty contributor is

$$u_x = ba$$

where b = the transformation coefficient given in Fig. 18.120

The rectangular distribution is to be taken when it is not sure that the normal distribution applies, and the U-distribution is unlikely. The U-distribution occurs with phenomena varying similar to a sine wave, e.g. many time dependent phenomena.

The expanded measurement uncertainty U is

$$U = ku_c$$

If not otherwise specified, according to ISO 14 253-1 the coverage factor $k = 2$ applies which corresponds to a confidence level of = 95%.

ISO TR 14 253-2 contains a list of possible measurement uncertainty contributors (u_x) to the combined measurement uncertainty u_c, e.g.:

- deviation of the calibration standard;
- deviation of the measuring equipment (e.g. maximum permissible error (MPE) of the measuring instrument);
- deviation caused by support and alignment of the workpiece (set-up);
- deviation caused by incomplete tracing of the workpiece (form deviation);
- deviation caused by temperature influences (deviations from 20°C);
- deviation caused by measuring force (deformations);
- deviation caused by gravity influences on the workpiece (deformations);
- deviation caused by the metrologist.

When the influences are known in units other than length units, e.g. temperature units, they must be transformed into length units.

The PUMA method is a step-by-step iterative procedure where, due to improved knowledge or due to improved measuring processes, the assessed measurement uncertainty is reduced.

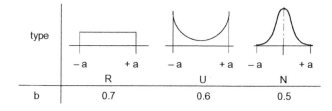

Fig. 18.120 Transformation coefficient b

In the first iteration the evaluation type B is used and thereby time- and cost-consuming experiments are avoided.

Due to the squares of the measurement uncertainty components u_x, components with small values of u_x contribute very little to the combined measurement uncertainty u_c. Therefore, if the measurement uncertainty shall be reduced, such contributors should be dealt with that have large values. ISO TR 14 253-2 describes a scheme for procedure and documentation of the assessment of the measurement uncertainty.

In the field of geometrical deviations the assessment of the measurement uncertainty is often very difficult. It is difficult to identify which contributors are effective and to what magnitude.

Therefore at present in the inspection of workpieces the measurement uncertainty is more or less roughly estimated according to experiences.

Means for a more precise assessment of the measurement uncertainty are in preparation as follows.

(a) Uncertainty budgeting according to ISO TR 14 253-2

It is intended to develop a collection of examples of the assessment of the measurement uncertainty for various methods of measurement of geometrical deviations, e.g. methods according to ISO TR 5460. The examples may be used as guidelines. (ISO 14 253-2 shows in Annex A one example of the estimation of the measurement uncertainty of a roundness measurement.)

It is also intended to develop a similar PC software that asks for the values of the contributors (e.g. temperatures, maximum permissible errors (MPE) of the measuring instruments, deviations of auxiliary equipment) and calculates the measurement uncertainty.

(b) Virtual coordinate measuring machine ISO 15 530*

For coordinate measuring machines (CMM), there has been developed a computer program that estimates the measurement uncertainty by simulation of repeated measurements.

For this program the uncorrected deviations of the CMM are assessed with the aid of a calibrated sphere and a calibrated sphere plate. The program simulates measurements at the points originally measured, taking into account the errors of the CMM. From these simulated measurements the program calculates the measurement uncertainty. After a single measurement the instrument gives the measurement result together with the measurement uncertainty.

(c) Comparison (substitution) method ISO 15 530*

With this method the workpiece and a similar calibrated object (e.g. a calibrated workpiece) are measured alternately. The measuring instrument serves as a comparator. The essential contributors to the measurement uncertainty are:

- contributors due to the measuring procedure;
- contributors due to the calibration;
- contributors due to the workpiece (inhomogeneities of form deviations, roughness and material properties).

* ISO Standard in preparation.

This procedure leads to relatively small measurement uncertainties. But the practical use is very limited because of the effort needed for calibration and for logistics of the calibrated objects.

Another method is the task-related calibration of the measuring process. The calibrated object is measured repeatedly and the measurement uncertainty of the measuring process is assessed. This value is a measurement uncertainty contributor for the measurement of similar workpieces. Also this method has limited practical use because of the effort needed for calibration and for logistics of the calibrated objects.

(d) Measurement of long dimensions and of other geometrical deviations

With the measurement of long dimensions, e.g. diameters, widths or positional toleranced dimensions, the influence of the temperature is often the largest or even dominant measurement uncertainty. Neumann (Ref. [9]) has found for measurements of workpieces of steel with CMMs the measurement uncertainty contributor due to measurements deviating $\pm 2\,K$ from 20°C:

when temperature has not been corrected: $\pm U_{95} \approx \pm 40\,\mu m/mK$,
when temperature has been corrected: $\pm U_{95} \approx \pm (5 \ldots 20)\,\mu m/mK$.[†]

When measuring deviations of form, orientation, coaxiality, symmetry and run-out, the measured value is close to zero. Therefore the influence of the temperature is much smaller and other influences are more important (e.g. MPE or calibration uncertainty of the measuring device and the repeatability of the measuring process).

18.11.4 Calibration of measuring instruments

Calibration is (according to GUM) a set of operations that establish under specified conditions the relationship between the value indicated by the measuring instrument and the corresponding value realized by standards.

Measurement standard is (according to GUM) material measure, measuring instrument, reference material or measuring system intended to define, realize, conserve or reproduce a unit or one or more values of a quantity, e.g. a standard for a length.

In order to assess the measurement uncertainty the measuring instrument must be calibrated and the calibration uncertainty must be taken into account. However, the calibration of metrological characteristics of measuring equipment that have no influence on the intended measurement is superfluous and uneconomic. For example, a length measuring instrument used only for measurements of a short range of lengths need not to be calibrated over the whole measuring range of the instrument.

---

* ISO Standard in preparation.
† Depending on the accuracy of the assessment of the temperature.

19

Function-, Manufacturing-, and Inspection-Related Geometrical Tolerancing

19.1 Definitions

19.1.1 Function-related geometrical dimensioning and tolerancing

Geometrical dimensioning and tolerancing (GD&T) are function related when the functional requirements are directly indicated and toleranced (Fig. 19.1). Then the largest tolerances appear.

Figure 19.2 shows extreme cases of the permissible workpiece shape according to GD&T of Fig. 19.1. A gauge of width 4 and centre distance 15 must fit into the workpiece.

19.1.2 Manufacturing-related geometrical dimensioning and tolerancing

GD&T are manufacturing related when the dimensions and tolerances that are to be adjusted during manufacturing are directly indicated (there is no need for the worker to calculate) and when these tolerances can be respected by the intended manufacturing process (Fig. 19.3). This is the easiest way to control the manufacturing process. However, the functional requirements, the function-related tolerances, must not be violated. If function-related GD&T and manufacturing-related GD&T do not coincide, the functional requirements are indicated indirectly. This leads to a reduction (decrease) in the tolerances.

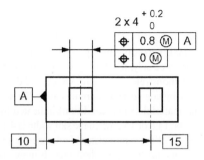

Fig. 19.1 Example of function-related D&T (plug fit)

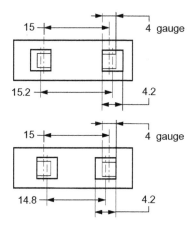

Fig. 19.2 Permissible extreme cases according to Fig. 19.1

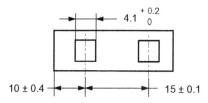

Fig. 19.3 Example of manufacturing related D&T

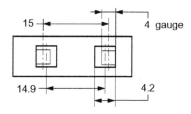

Fig. 19.4 Derivation of the least centre distance (14.9) from Fig. 19.1

For example, according to Fig. 19.3, the centre distance must not exceed 15.1, but with function-related GD&T, according to Fig. 19.1, the centre distance may be 15.2 when the actual size (more precisely, the mating size) of the slot is 4.2.

Figures 19.4 and 19.5 demonstrate how the tolerances are derived from the function-related GD&T of Fig. 19.1 by considering extreme cases. The workpiece must not violate the limits (maximum material virtual condition) indicated in Fig. 19.1.

The centre distance may be 14.8 or 15.2 when the slot width is 4.1. Therefore this GD&T Fig. 19.3, leads to a reduction of the centre distance tolerance compared with Fig. 19.1.

Often, depending on the manufacturing method, there is more than one manufacturing-related GD&T. The manufacturing-related GD&T according to Fig. 19.3 applies to a

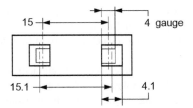

Fig. 19.5 Derivation of the maximum centre distance (15.1) from Fig. 19.1

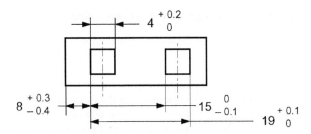

Fig. 19.6 Example of manufacturing-related D&T

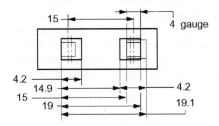

Fig. 19.7 Derivation of the limits of size (15 − 0.1 and 19 + 0.1) from Fig. 19.1

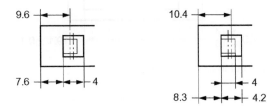

Fig. 19.8 Derivation of the limits of size of the edge distance from Fig. 19.1

manufacturing method where the slot centres are to be adjusted. When during manufacture the slot faces are directly adjusted, the manufacturing-related GD&T could be according to Fig. 19.6.

Figures 19.7 and 19.8 demonstrate how the tolerances are derived from the function-related GD&T of Fig. 19.1 considering extreme cases. The workpiece must not violate the limits (maximum material virtual condition) indicated in Fig. 19.1.

The distance of the left slot sides may be 14.8 when the distance of the outer slot sides is 19. However, the distance of 14.8 is not permissible when the distance of the outer slot sides is 19.1. Therefore this GD&T (Fig. 19.6) leads to a reduction in the tolerances compared with Fig. 19.1.

It is also possible to indicate and tolerance the manufacturing process variables (e.g. substitute size and tolerance, form tolerance, substitute distance and substitute location tolerance). Statistical tolerancing may be also advantageous. However, appropriate standards are not yet available, see 8 and 14.

19.1.3 Inspection-related geometrical dimensioning and tolerancing

GD&T is inspection related when the dimensions and tolerances to be inspected are directly indicated and when the tolerances (deviations) can be inspected with sufficient accuracy. The inspection-related GD&T depends on the inspection method. When the maximum material requirement applies and an appropriate gauge is available, the inspection-related GD&T coincides with the function-related GD&T. If the inspection uses measurement techniques, the inspection-related GD&T may correspond to Fig. 19.9. There the total tolerance (Fig. 19.1) is split into tolerances of the slot width and of the slot side distances, which leads to a reduction in the slot width tolerances compared with Fig. 19.1.

Figures 19.10 to 19.12 demonstrate how the tolerances are derived from the function-related GD&T of Fig. 19.1 considering extreme cases. The workpiece must not violate the limits (maximum material virtual condition) indicated in Fig. 19.1.

The permissible inspection size may be 19.4 when the slot width is 4.2. However, the inspection size of 19.4 is not permissible when the slot width is 4.1. Therefore this GD&T (Fig. 19.9) leads to a reduction in the tolerances compared with Fig. 19.1.

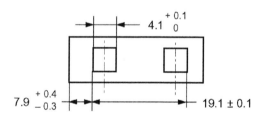

Fig. 19.9 Example of inspection related D&T

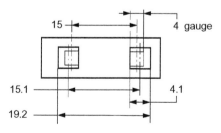

Fig. 19.10 Derivation of the minimum inspection size (19.0) from Fig. 19.1

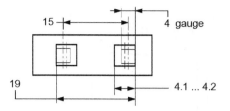

Fig. 19.11 Derivation of the maximum inspection size (19.2) from Fig. 19.1

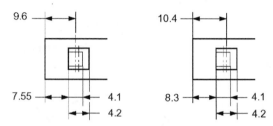

Fig. 19.12 Derivation of the inspection sizes of the edge distances from Fig. 19.1

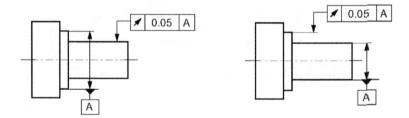

Fig. 19.13 Tolerancing: inspection appropriate (right) and not inspection appropriate (left)

Indications are not inspection appropriate when inspection is not possible with sufficient accuracy (i.e. when the measurement uncertainty becomes too large). For example, tolerancing according to Fig. 19.13 left is not inspection appropriate, because the datum is too short.

If necessary, a further datum is to be indicated (e.g. the flange face) in order to stabilize the orientation of the workpiece (Fig. 19.14).

Figure 19.15 shows non-inspection appropriate tolerancing. The measurement uncertainty of the location of the axis derived from a part of a cylinder smaller than half of the cylinder is very large (often larger than the tolerance).

Figure 19.16 shows tolerancing that is inspection appropriate. The deviation of the surface from the theoretical exact form at the specified location can be measured with a smaller measurement uncertainty.

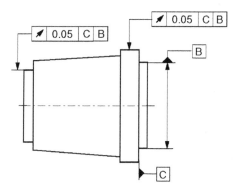

Fig. 19.14 Short datum and further datum in order to stabilize the orientation of the workpiece

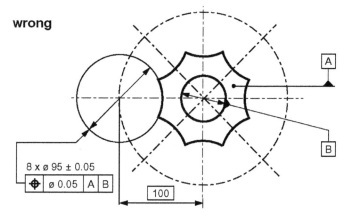

Fig. 19.15 Non-inspection appropriate tolerancing

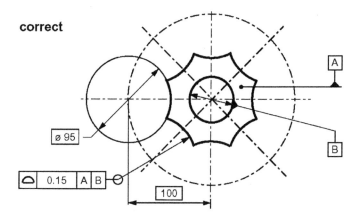

Fig. 19.16 Inspection appropriate tolerancing

19.2 Method of geometrical dimensioning and tolerancing

The function-related geometrical dimensioning and tolerancing has the advantage of indicating

- the largest tolerances;
- the limits that must not be violated by any combination of variables;
- the limits from which all other methods of GD&T are derived.

Because of these advantages, function-related GD&T should always be provided. Sometimes (e.g. with mass production) further drawings, manufacturing plans or inspection plans with manufacturing-related or inspection-related GD&T derived from the function-related GD&T could be helpful. CAD provides the possibility of storing the function-related GD&T, the manufacturing-related GD&T and the inspection-related GD&T on different layers of the same drawing data.

Sometimes general guides can be helpful indicating in what functional cases and in what ways manufacturing-related or inspection-related D&T can be derived from the function-related D&T. Such guides can, for example, specify that in cases where coaxiality tolerances are indicated but no suitable form measuring instruments or coordinate measuring machines are available, the radial run-out tolerance of the same value shall apply. The same applies in cases where coaxiality tolerances with the maximum material requirement are indicated and no appropriate gauges are available. Then only in cases of dispute need the more precise and more expensive inspection method of the coaxiality tolerance be applied.

Geometrical tolerances specify geometrical tolerance zones. Particular manufacturing or inspection methods are not specified by geometrical tolerances (however, the tolerances must be respected and verified by the chosen method). Whether inspection is necessary, and to which level and by which method, depends on:

- level of control of the manufacturing process;
- probability of occurrence of deviations (e.g. of lobed forms);
- consequences of exceeding the tolerance;
- confidence in the manufacturer (e.g. inspection history).

Specification of particular inspection methods would force manufacturers to provide particular inspection devices, in addition to those already possessed for geometrical tolerancing, thus leading to permanent expansion in instrumentation. Further, there is often a contradiction between the geometrical definition of the geometrical tolerance and the inspection method. Therefore it would be necessary to specify all details of the inspection methods, which would lead to a multiplication of inspection methods and would be detrimental to technical advance. Therefore ISO 1101 does not provide a drawing indication of the inspection method.

Proper application of GPS GD&T needs education of the staff. The software described in 19.4 may be helpful.

19.3 Assessment of function-related geometrical dimensioning and tolerancing

The following lists give hints as to how to proceed in assessing the function-related geometrical tolerancing.

19.3.1 Tolerancing of orientation, location and run-out

- Which features are related to each other by the function?
- Is the general geometrical tolerance sufficient (e.g. ISO 2768-2)? If it is not then
- Does one feature determine orientation or location and may therefore serve as a datum?
- Is it appropriate to specify a common tolerance zone and thereby avoid the specification of a datum?
- Is it appropriate to specify a datum system or datum targets?
- Which characteristics shall be toleranced?
- Is Ⓜ, Ⓛ, Ⓡ, Ⓟ or Ⓕ appropriate?
- What is the magnitude of the tolerance?

19.3.2 Tolerancing of form

- Which characteristic shall be toleranced?
- Is Ⓔ appropriate and sufficient? If it is not then
- Is the general geometrical tolerance sufficient (e.g. ISO 2768-2)? If it is not then
- Is the form deviation already limited sufficiently by a tolerance of orientation or location or total run-out? If it is not then
- What is the magnitude of the tolerance?

19.3.3 Type of tolerance

- Fits: Ⓔ.
- Clearance fits, but no kinematics: Ⓜ (then often Ⓔ is not needed, see 20.7.3).
- Geometrical ideal form within the material required: Ⓛ.
- Weight limitation: thickness tolerance (maximum thickness).
- Interference fit, kinematics, optics, electrical contacts, measuring contacts: geometrical tolerances regardless of other tolerances.
- Threaded holes, holes for centre pins: Ⓟ.
- Flexible parts: Ⓕ, ISO 10579-NR.

Often (e.g. with clearance fits associated with the maximum material requirement Ⓜ) the features are of the same functional priority with regard to being chosen as a datum. Then each of them may be chosen as a datum (Figs 20.47 to 20.50). However, the dependence shown in Fig. 18.9 is to be observed. In contrast, in case of interference fits, according to Fig. 20.95, the assembly direction determines the datum.

19.4 Assessment of the optimum geometrical dimensioning and tolerancing

Function-related GD&T depends only on the design. However, some designs provide larger and more economical tolerances than others for the same functional purpose. A value analysis provides help to recognize this.

Whether in addition to the function-related GD&T other specifications should be issued providing manufacturing- and/or inspection-related GD&T, or whether the drawing should be modified, depends on economic considerations. Participants in these economic considerations should be all concerned departments.

The **design department** is responsible for the tolerances guaranteeing functionality.

The **manufacturing planning department** is responsible for the tolerances being achievable.

The **manufacturing department** is responsible for the tolerances being respected.

The **inspection planning department** is responsible for the tolerances allowing appropriate inspection.

The **inspection department** is responsible for the tolerances being inspected.

The **standardization department** is responsible for all concerned receiving the necessary information on standards and their applications.

With the cooperation of these departments, the optimum GD&T should be elaborated for typical parts, which afterwards become models for the design departments. Experience has shown that without such cooperation no economical GD&T can be achieved.

There is a **software** available for company intranets and for single users as a help for correct GD&T. It shows the symbols and definitions of GD&T, gives the rules for the application and shows examples of GD&T for many functional cases. The differences between ISO and ASME are explained and a guide for serving both systems simultaneously is included.

The software serves as a help for reading drawings and for establishing correct drawings.

The software can be ordered as follows:

G. Henzold: Examples of GD&T, sales no. 79 526
(DIN) Beuth-Verlag GmbH D-10772 Berlin
Fax: +49-30-2601-18 01, E-mail: foreignsales@beuth.de

20

Examples of Geometrical Tolerancing*†

20.1 Restrictions of geometrical tolerances

20.1.1 Restricted range of application

Figures 20.1 and 20.2 show examples of restricted ranges of application of geometrical tolerances.

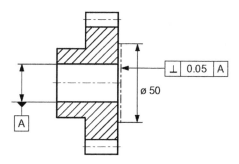

Fig. 20.1 Restricted application of perpendicularity tolerance

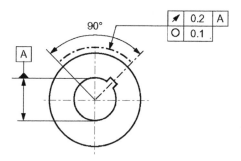

Fig. 20.2 Restricted application of run-out and roundness tolerance

* See also 19.4 (previous page).
† Further examples are described by Lowell W. Foster (Ref. [14]).

20.1.2 Geometrical tolerances and additional smaller tolerances of their components

It may be that, besides the geometrical tolerances, their components may have to have narrower tolerances. For example, journal shafts may require smaller tolerances for roundness than for cylindricity in order to maintain proper lubrication (Fig. 20.3).

According to Fig. 20.4, the surface of the right cylinder may vary between two cylinders coaxial with the datum A. According to Fig. 20.5, the parallelism deviation

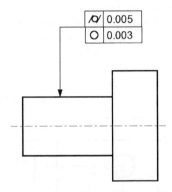

Fig. 20.3 Cylindricity tolerance and smaller tolerance of its component roundness

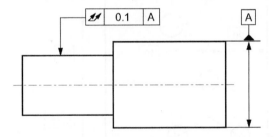

Fig. 20.4 Total run-out tolerance

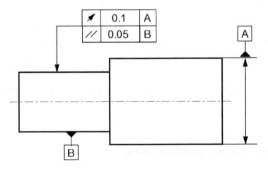

Fig. 20.5 Run-out tolerance and smaller tolerance of parallelism of the generator lines

(and the straightness deviation) of the generator lines are restricted to a smaller tolerance (0.05).

20.1.3 Superposition of positional tolerances

Figure 20.6 shows two groups of holes. The four hole group is related to the datums D, A and B. The maximum material virtual conditions (ø7.85) of this group together establish the datum C of the positional tolerances ø0.15 of the three hole group. That is, perpendicular to D, the maximum material virtual condition of the four holes (ø7.85) together with the maximum material conditions of the three holes (ø5.75) must fit (simultaneously) into the holes. In addition, perpendicular to D, the maximum material virtual conditions (ø5.85) for the positional tolerance ø0.05 of the three holes must fit together into the three holes.

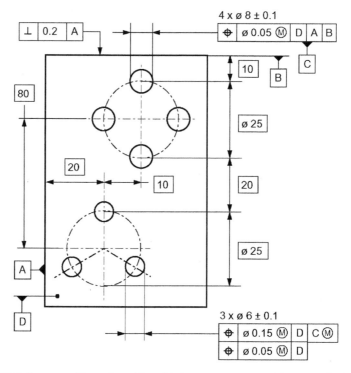

Fig. 20.6 Superposition of positional tolerances; group of features (holes) as datum

20.2 Tolerances for section lines

When flatness tolerances are indicated, the maximum permissible deviation in straightness is equal in each direction (Fig. 20.7 (a)). However, sometimes (e.g. in the case of slideways)

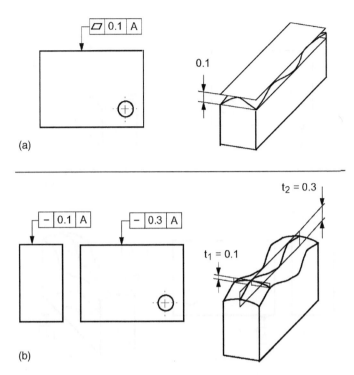

Fig. 20.7 Flatness tolerance (a) and different straightness tolerances (b) of the same surface

the form deviation may be greater in one direction than in the other, Fig. 20.7 (b) shows an example.

Parallelism tolerances of surfaces define three-dimensional tolerance zones in which the surface must be contained (Fig. 20.8 (a)). When (e.g. for slideways) the parallelism deviation in one direction may be larger than in the other, the parallelism tolerance has to be applied to section lines (Fig. 20.8 (b)).

The indication "lines" does not occur when a line profile tolerance (symbol ⌒) is indicated instead of the parallelism tolerance (Fig. 20.8(d)).

The datum remains a plane in all cases of Figs 20.7 (a) and (b) and 20.8 (a) and (b). When the datum shall become a cross-section line the indication "lines" is needed (Fig. 20.8 (c) and (d).

The indication "lines" is not necessary when the drawing shows that only lines can possibly be applied (Fig. 20.9) (if necessary the indication "high point line" or "section line" is to be indicated, see 3.3.1).

20.3 Tolerances of profiles

With curved profiles (e.g. of turbine blades) the profile tolerance sometimes varies steadily along the curve. Figure 20.10 shows the drawing indication and the relevant

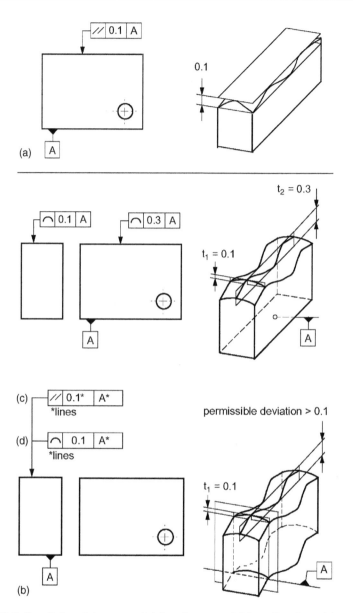

Fig. 20.8 Parallelism tolerance of: (a) surfaces and (b) section lines (c) and (d) Parallelism tolerance of lines relative to lines

tolerance zone. The curvature of the tolerance zone boundary between the indicated points is not standardized, but usually splines are used.

Figure 20.11 shows a profile tolerance of a curved tube. To limit buckling (e.g. that caused by the supports during the bending procedure), the profile tolerance for a limited length and the roundness tolerance are indicated.

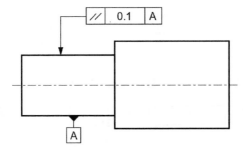

Fig. 20.9 Parallelism tolerance of lines of a cylinder

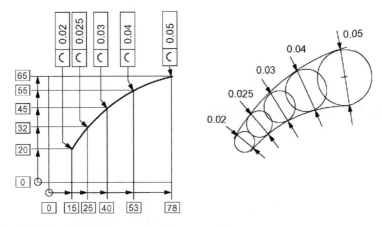

Fig. 20.10 Profile tolerance of varying magnitude

Figure 20.12 shows an example of different tolerance zones along a surface. The form of transition from one zone to the next is not standardized, and may require the indication of a note (e.g. "tangential transition").

Figure 20.13 shows an example of different tolerances for form, orientation and location of the same feature (window). The TED 300 does not apply to the orientation tolerance 1.

20.4 Position of a plane

The indications (a) and (b) in Fig. 20.14 are permissible and define the same tolerance zone. As not only the orientation and form (flatness) but also the location (related to datum B) is important, the indication (c) (orientation tolerance) is not appropriate. ISO 1101 uses the indication (a) or (b), ASME Y14.5 uses the indication (b).

In Fig. 20.15 the actual local sizes are limited by the surface profile tolerance of one surface and the flatness tolerance of the other surface. The actual local sizes may vary between 49.85 and 50.05.

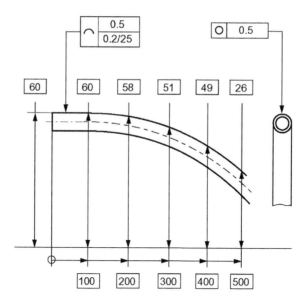

Fig. 20.11 Tolerancing to limit buckling

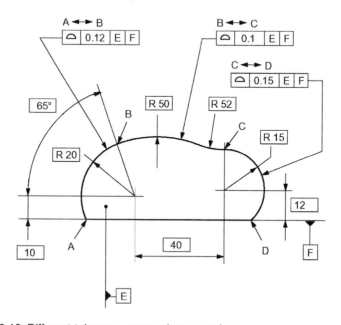

Fig. 20.12 Different tolerance zones along a surface

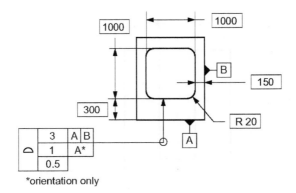

Fig. 20.13 Different tolerances for form, orientation and location of the window

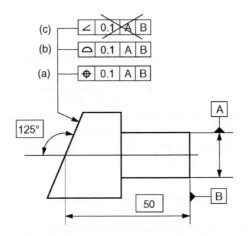

Fig. 20.14 Tolerancing of the position of a planar surface

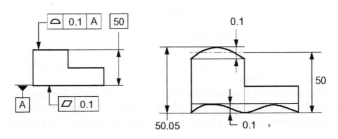

Fig. 20.15 Tolerancing of the position of a planar surface parallel to a datum plane with a flatness tolerance to the datum surface

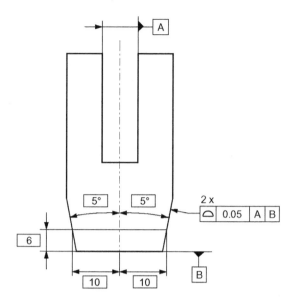

Fig. 20.16 Positional tolerancing of two inclined planar surfaces (wedge)

Figure 20.16 shows the tolerancing of the position of two inclined planar surfaces. The theoretical exact position is determined by the datum A (median plane of the slot), the inclination (5°) and the position of the lines (distance 10 from datum A and distance 6 from datum B). The surface profile tolerances also limit the symmetry deviation of the wedge. The eccentricity (distance of the median face of the wedge from the datum median plane A) can vary up to 0.025.

20.5 Perpendicularity tolerances in different combinations

Figure 20.17 (top) shows a perpendicularity tolerance and the permissible workpiece (e.g. an inclined workpiece). Figure 20.17 (bottom) shows the tolerancing limiting all perpendicularity deviations.

20.6 Location of axes and median faces

20.6.1 Angular location of features

In Fig. 20.18(a) the keyways are drawn cosymmetrically. The symmetry deviation is limited by the general tolerance.

This limitation has no economical disadvantage, even when the function allows a random angular location, because the general tolerance is respected anyway without any additional effort (see 16.2). However, if the angular location is being optional, this may be indicated in the drawing (Fig. 20.18(b)).

Figure 20.19 above shows tolerancing of the angular location of the upper hole. Angular dimensional tolerancing according to (a) is wrong because it is not clear where

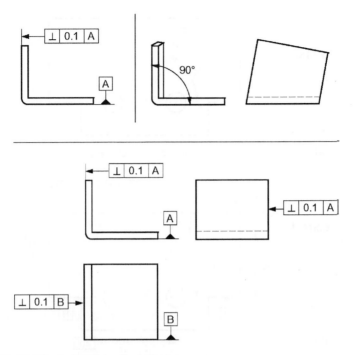

Fig. 20.17 Effect of perpendicularity tolerances

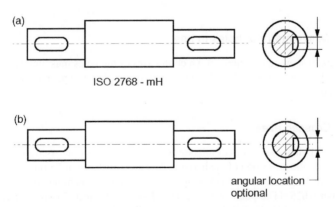

Fig. 20.18 Cosymmetrically drawn keyways: (a) symmetrical according to ISO 2768-mH, (b) with angular locations optional (also between each other)

the apex of the angle is located (related to the holes, to the two horizontal planes, to the median plane of the two upper vertical planes, to the two lower vertical planes or to the two holes).

Figure 20.19 (b) shows correct positional tolerancing with theoretical exact location of the angle apex.

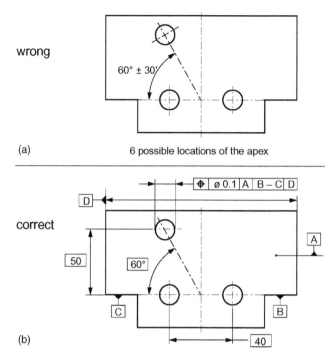

Fig. 20.19 Angular location of a hole axis

20.6.2 Cosymmetry of holes in line

Figure 20.20 shows an example in which the axes of the holes must not deviate more than ±0.05 from a plane. The distances of the axes to the upper plane may vary within ±0.5.

20.6.3 Crossed axes

Figure 20.21 shows examples of tolerances for the distances of crossed axes.

In case (a) the projection of the horizontal axis at a distance of 75 (i.e. for this point and over the length zero) must not be apart from the projection of the datum axis A for more than 0.025. (The point of the projected axis at distance 75 must be contained between two parallel planes 0.05 apart; their median plane contains the datum axis A.) This tolerance does not limit the angle between the two axes.

In case (b) the projection of the horizontal axis must be contained in a tolerance cylinder of ø0.05 over a length of 75. The axis of the tolerance cylinder meets the datum axis A; the two axes form an angle of 90°. In this case not only the distance between the workpiece axes is toleranced but also the deviation from the 90° angle between them.

Figure 20.22 shows two different indications (a) and (b) for tolerancing the perpendicularity deviation of two holes. The indication (a) or (b) alone does not limit the deviation

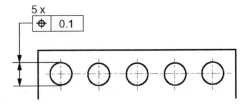

Fig. 20.20 Cosymmetry of holes in line

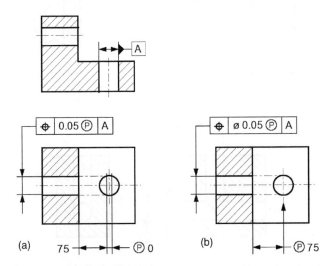

Fig. 20.21 Tolerancing of crossed axes

perpendicular to the projection plane (eccentricity). The additional indication (c) limits the eccentricity.

The indication in Fig. 20.22 (below) limits the eccentricity of the axes of the hole in relation to the datum axis A.

20.6.4 Different locational tolerances for different features drawn on the same centre line

When on a drawing features are drawn on the same centre line, it is difficult to allocate different distance tolerances to the different features. For example, in Fig. 20.23 the indication applies for the holes as well as for the slots. Positional tolerancing, however, provides the possibility of allocating different tolerances to the different features (Figs 20.24 and 20.25).

In Fig. 20.26 holes and countersinks have different positional tolerances with different datums. The datum for the positional tolerance of the countersink is the relevant hole axis.

The positional tolerances according to Fig. 20.27 allow positional tolerancing of the median faces of the slotted holes in relation to the median plane A. With the indication 10 ± 0.1 instead of the positional tolerances (Fig. 20.28) the location of the slotted holes depends on the actual locations of the two cylindrical holes and should be avoided.

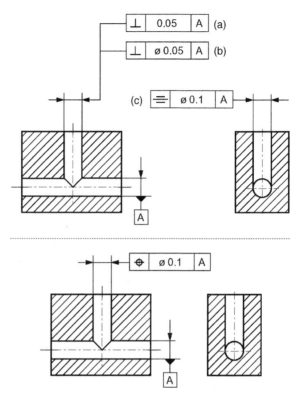

Fig. 20.22 Tolerancing for axes of holes; perpendicularity tolerancing for axes of holes (a) and (b) alone do not limit the eccentricity but the additional indication (c) does; positional tolerancing (below) limits perpendicularity and eccentricity

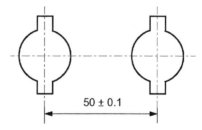

Fig. 20.23 Distance tolerance of features drawn on the same centre line

20.6.5 Positional tolerances for features located symmetrically to a symmetry line and multiple patterns of features

ISO 5458 states that positional toleranced features drawn on the same centre line are regarded as related features having the same theoretical exact location (see 6.2). However, sometimes (e.g. when the features are drawn not on but symmetrically with respect to

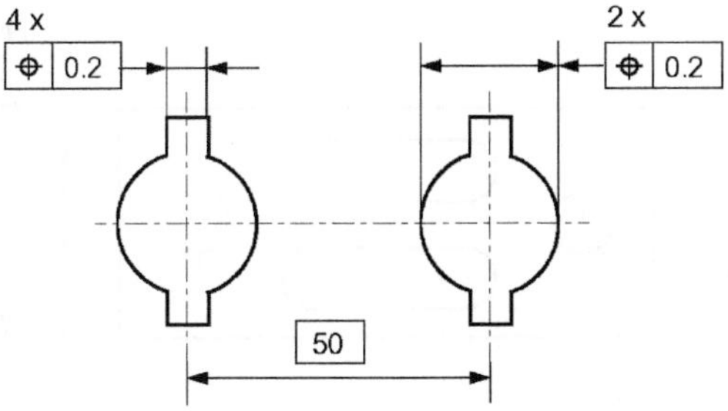

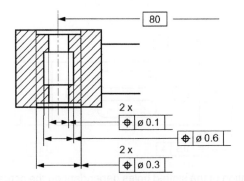

Fig. 20.24 Features drawn on the same centre line, but with different positional tolerances

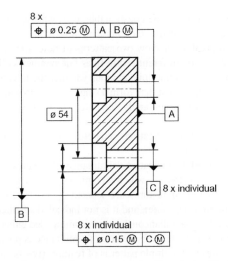

Fig. 20.25 Features drawn on the same centre line, but with different positional tolerances

Fig. 20.26 Features drawn on the same centre line, but with different positional tolerances related to different datums

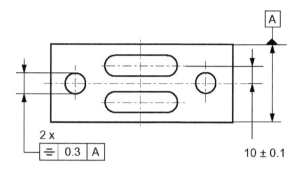

Fig. 20.27 Differentiation of locational tolerances in relation to the same median plane by positional tolerances (recommended)

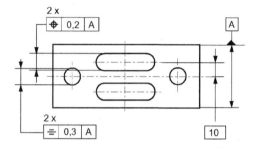

Fig. 20.28 Location of the slotted holes dependent on the actual location of the cylindrical holes, toleranced by ±tolerances (not recommended)

a common centre line), it might be questionable as to what applies. In such cases an appropriate note should be added to the tolerance indication.

Figures 20.29 and 20.30 each show two patterns of holes (ø5 and ø8). The features of the patterns are drawn on different centre lines but symmetrically with respect to a symmetry line. From this it is not quite obvious whether the tolerances apply simultaneously to the two patterns so that they form a single pattern to which a common gauge applies or whether the tolerances apply separately to each pattern so that separate gauges apply to the patterns.

In the case of Fig. 20.29 the indication "separate" allows both patterns of holes to be regarded as separate patterns.

In the case of Fig. 20.30 the indication "simultaneous" ensures that both patterns are to be regarded as a single pattern to which a common gauge applies.

When positional toleranced features are located symmetrically with respect to a symmetry line (common centre line), and it is not indicated whether the positional tolerances apply separately or simultaneously, it is recommended either to ask the designer what shall apply or to use the failsafe method of simultaneous positional tolerances.

ASME Y14.5M states that multiple patterns of features (two or more patterns of features not necessarily located symmetrically to each other) located by theoretical exact dimensions relative to the same datum system (same datum features referenced in the same order of precedence and in the same modification, i.e. with the same modifier or without modifier) are to be regarded as a single pattern (simultaneous requirements). If this is not required this should be indicated by "SEP REQT" under each tolerance frame (instead of "separate" in Fig. 20.29.

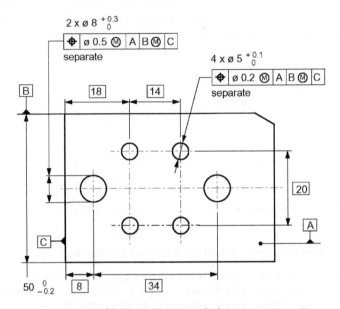

Fig. 20.29 Two patterns of holes to be regarded as separate patterns

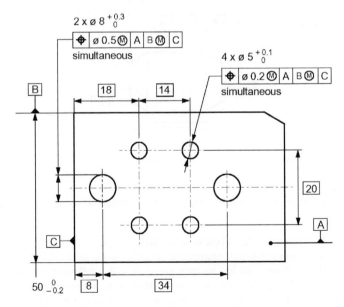

Fig. 20.30 Two patterns of holes to be regarded as a single (common) pattern

20.7 Datums

20.7.1 Plane surface as datum for a symmetry tolerance

Normally the datum for a symmetry tolerance is a median plane or an axis. In special cases the datum may be a plane surface. Then the simulated datum feature has to be positioned according to the minimum rock requirement (Fig. 20.31).

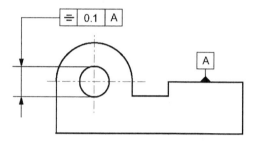

Fig. 20.31 Plane surface as datum for a symmetry tolerance

20.7.2 Plane surface as datum for a run-out tolerance

Normally the datum for a run-out tolerance is an axis. With machines (e.g. electric motors) it might happen that the datum axis cannot be detected. In such cases the housing base or housing flange may be indicated as the datum (Fig. 20.32).

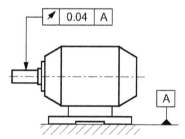

Fig. 20.32 Plane surface as datum for a run-out tolerance

20.7.3 Envelope requirement and maximum material requirement with datums

Figure 20.33 shows on the left a hole to which the envelope requirement Ⓔ applies. On the right the maximum material requirement Ⓜ applies to the datum hole A. This datum hole is a primary datum, and has no form tolerance with a following symbol Ⓜ. This (for the datum hole) has the same effect as the envelope requirement (it must be possible to fit a cylindrical gauge of maximum material size). Therefore in tolerancing the hole on the right, the symbol Ⓔ is omitted.

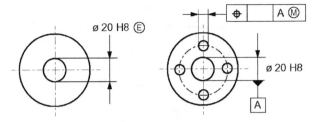

Fig. 20.33 Envelope requirement and datum with maximum material requirement

Note that ISO 2692 - 1988 contains on page 20, Fig. 29a, the symbol Ⓔ behind the size tolerance of the datum, although at the datum the maximum material requirement Ⓜ is indicated. The indication Ⓜ, however, implies the requirement of Ⓔ. In order to avoid duplication of requirement indications, Ⓔ should be omitted or enclosed in parentheses.

20.7.4 Positional tolerances with and without datums

Positional tolerances may be applied with or without datums. Without a datum, all features combined by the positional tolerance have the same positional tolerance (Fig. 20.34).

When one hole is indicated as a datum (Fig. 20.35), this hole has no positional tolerance in relation to the other hole(s). According to Fig. 20.35, the gauge has maximum material size = ø7.8 at the datum and maximum material size minus positional tolerance = ø7.7 at the other feature. According to Fig. 20.34 the gauge has ø7.7 at both features.

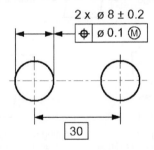

Fig. 20.34 Positional tolerances of a group of holes without datum

20.7.5 Sequence and maximum material requirement for datums

Figure 20.36 shows different possibilities for datums.

In case (a) the surface B must contact the gauge according to the minimum rock requirement. The diameter of the gauge is equal to the maximum material size of the datum feature A.

In case (b) the gauge must enclose the datum feature A with the least possible diameter (mating size of datum feature A). The datum feature is orientated according to the minimum rock requirement. The surface B contacts the gauge on one point only. (In the case of the positional tolerance of the holes this has no influence on the position.)

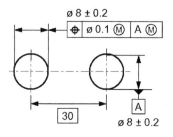

Fig. 20.35 Positional tolerances of a group of holes with datum hole

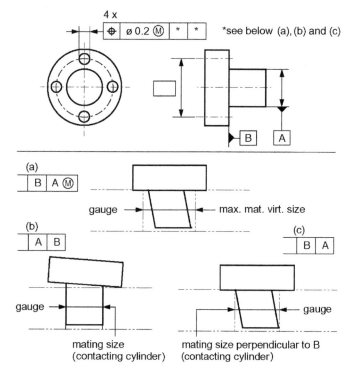

Fig. 20.36 Sequence and maximum material requirement for datums

In case (c) the surface B must contact the gauge according to the minimum rock requirement. In this orientation the gauge must enclose the datum feature A with the least possible diameter.

Figures 20.37 and 20.38 show positional toleranced holes with and without application of the maximum material requirement for the datums.

According to Fig. 20.37 the surface A must contact the gauge according to the minimum rock requirement. The maximum possible cylinder perpendicular to the gauge surface A must enclose the datum feature B. The slot C must be enclosed by the maximum possible parallelepiped perpendicular to the gauge surface A. The median plane

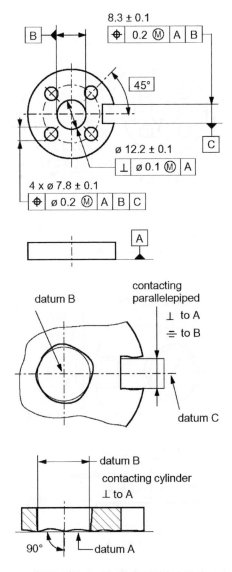

Fig. 20.37 Datums without maximum material requirement

of the parallelepiped contains the axes of the gauge cylinder B. The maximum material virtual conditions of the four holes (gauge) are ø7.5.

According to Fig. 20.38 also, surface A must contact the gauge according to the minimum rock requirement. The gauge cylinder B and the gauge parallelepiped C are perpendicular to the gauge surface A, and the symmetry plane of the gauge parallelepiped contains the axis of the gauge cylinder B. However, the gauge cylinder B and gauge

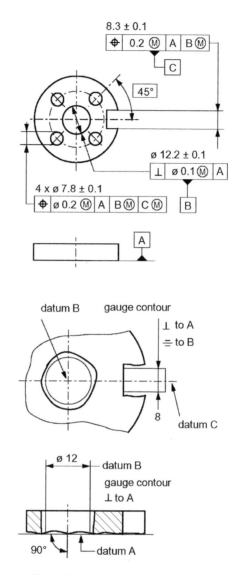

Fig. 20.38 Datums with maximum material requirement

parallelepiped C have maximum material virtual size (ø12 and 8). The maximum material virtual conditions of the four holes (gauge) are ø7.5.

20.7.6 Interchanging toleranced feature and datum feature

Depending on how the datum feature and toleranced feature are chosen, different properties are toleranced. Tolerancing according to Fig. 20.39 does not, according to Fig. 20.40,

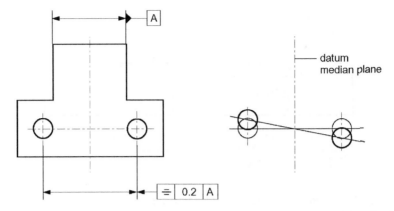

Fig. 20.39 Symmetry tolerance and permissible location of the holes; the perpendicularity deviation of the plane of the hole axes in relation to the median face of the tab is not limited

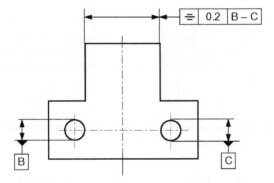

Fig. 20.40 Symmetry tolerance also limits the perpendicularity deviation of the symmetry face of the tab in relation to the plane of the hole axes

limit the perpendicularity deviation of the median face of the tab in relation to the plane of the hole axes.

See also 18.4.

20.8 Clearance fit

20.8.1 Coaxiality of holes

Figure 20.41 shows indications of flatness tolerances according to ISO 1101 (a) independent from each other (the tolerance zones may be different in height level and in orientation) and (b) as common tolerance zone for all two surfaces.

Figure 20.42 shows similar indications for the straightness of hole axes. It is evident that only the common zone indication limits the coaxiality deviation.

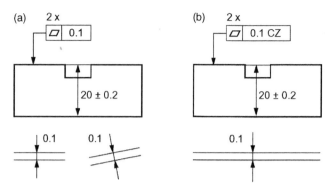

Fig. 20.41 Flatness tolerance zone: (a) independent from each other and (b) common zone

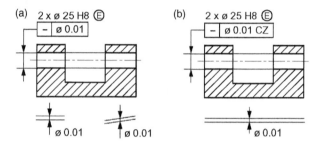

Fig. 20.42 Tolerance zone for the straightness of the axes: (a) independent from each other and (b) common zone

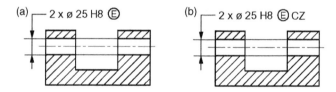

Fig. 20.43 Envelope requirement for holes drawn on the same centre line: (a) independent and (b) common

Figure 20.43 shows similar indications for the envelope requirement of holes. Only the indication "CZ" expresses the requirement that both actual holes must not infringe the imaginary envelope cylinder (boundary) of maximum material size that extends over both holes. Without the indication "CZ" (Fig. 20.43(a)), each hole must separately respect its envelope cylinder. These envelope cylinders need not be coaxial.

Figures 20.44 and 20.45 show the influence on the measured value by choosing the datum. On the right in both figures the measured value on the same actual workpiece is shown. The figures reveal that, on interchanging datum feature and toleranced feature on the same actual workpiece, quite different measured values may arise.

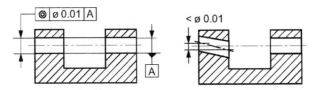

Fig. 20.44 Coaxiality tolerance and measured value

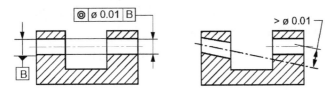

Fig. 20.45 Coaxiality tolerance and measured value with interchanged datum and toleranced feature compared with Fig. 20.44

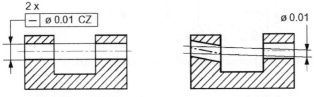

(According to ISO 286: ø25 H8 = ø25 + 0.033)

Fig. 20.46 Tolerancing of a clearance fit with common pin

Figure 20.46 shows tolerancing of the axes without datum. Both axes must be contained in the common tolerance zone ø0.01. Another possibility (Fig. 20.47) is that the tolerance corresponds to gauging with a plug gauge that extends over both holes simultaneously, the diameter of the plug gauge being maximum material size minus coaxiality tolerance (maximum material virtual size = ø24.99) in cases (a) to (c).

Figure 20.47 shows the possibilities for tolerancing of coaxial holes using coaxiality tolerances (According to ISO 1101 positional tolerances could also have been used with the same meaning.)

In case (a) each hole must respect the envelope requirement Ⓔ separately. The actual axis of the right hole must be contained in a tolerance cylinder of ø0.01 that is coaxial with the datum axis A. The datum axis A is positioned according to the minimum rock requirement. The right hole may have a coaxiality deviation of 0.005 (corresponding to a coaxiality tolerance of 0.01; see 18.7).

In case (b) each hole must respect the envelope requirement Ⓔ separately. The actual right hole must respect the maximum material virtual cylinder (gauge) of diameter given by the maximum material size minus the coaxiality tolerance (ø25 −0.01ø = ø24.99) that is coaxial with the datum axis A. The datum axis A is positioned according to the minimum rock requirement.

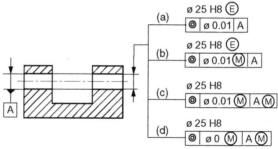

(According to ISO 286: ø25 H8 = ø25 + 0.033)

Fig. 20.47 Coaxial holes: possibilities of indicating coaxiality tolerances

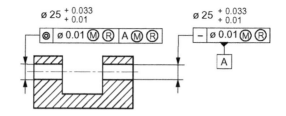

Fig. 20.48 Coaxial holes: reciprocity requirement applied

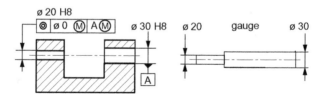

Fig. 20.49 Coaxiality of holes with different nominal sizes: tolerancing of clearance fit and relevant gauge

In case (c) a stepped plug gauge must fit in the holes, whose diameters are given by (right) the maximum material size minus the coaxiality tolerance (ø25 − 0.01 = ø24.99) and (left) the maximum material size (25).

In case (d) a plug gauge must fit simultaneously (coaxial) in both holes, whose diameter is the maximum material size (25).

Figure 20.48 shows a case similar to Fig. 20.47 (d); however, the total tolerance (0.033) is distributed on a size tolerance (0.023) and a geometrical tolerance (0.01) as a recommendation to the manufacturer. The requirement is for size and geometry to respect the maximum material virtual condition of ø25, i.e. a plug gauge of ø25 must fit simultaneously in both holes.

Figure 20.49 shows the tolerance indication and the relevant gauge when the holes have different nominal sizes.

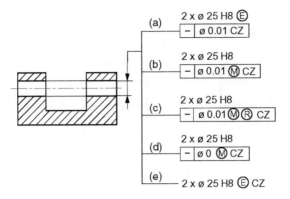

(According to ISO 286: ø25 H8 = ø25 + 0.033/−0)

Fig. 20.50 Coaxiality of holes: possibilities for indicating common zones

Figure 20.50 shows the possibilities for tolerancing coaxial holes using common tolerance zones "CZ" for the straightness deviations of the axes of the holes or common zones "CZ" for the envelope requirement.

In case (a) each hole must respect the envelope requirement Ⓔ separately. The actual axes of the holes must be contained within a common tolerance cylinder of ø0.01. The holes may have a coaxiality deviation of 0.01 (corresponding to a common straightness tolerance zone of ø0.01) regardless of the holes' sizes, i.e. also when the holes have maximum material sizes (ø25).

In case (b) a plug gauge must fit simultaneously (coaxially) into both holes, whose diameter is given by the maximum material size minus the straightness tolerance (ø25 − 0.01 = ø24.99). However, the actual local sizes of the holes must be between ø25 and ø25.033 (= ø25 H8).

In case (c) a plug gauge must fit simultaneously (coaxially) into both holes, whose diameter is given by the maximum material size minus the straightness tolerance (ø25 − 0.01 = ø24.99). The actual local sizes of the holes must be between ø24.99 and ø25.033.

In case (d) a plug gauge must fit simultaneously (coaxially) into both holes, whose diameter is the maximum material size (ø25).

In case (e) the same applies as in case (d).

Figure 20.51 shows a hinge frame with holes ø8 D 10 (+0.098/+0.040) that should fit together with a bolt ø8 h 9 (0/−0.036) without interference (clearance fit). Figure 20.51 (bottom) shows the extreme case when all actual sizes are at the worst limit. Even then, the parts can still fit without interference.

The holes may be inspected with a plug gauge whose diameter is given by the maximum material size minus the straightness tolerance (ø8.04 − 0.04 = ø8). The plug gauge must extend over both holes simultaneously.

Figures 20.52 to 20.55 show various possibilities for tolerancing coaxial holes in relation to a supporting surface (e.g. hinge). On the right the relevant inspection or gauge is shown.

Figure 20.56 shows the tolerancing of the gap of the hinge to fit with the tab. The maximum material virtual condition is perpendicular to the datums A and B (minimum rock requirement).

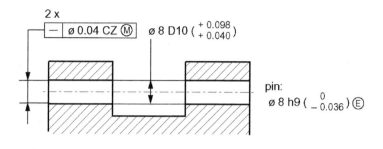

assembly without interference
extreme case: max. mat. condition

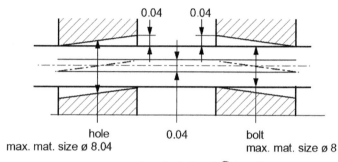

gauge contour for holes (Ⓜ) : ø 8

(According to ISO 286: ø8 D10 = ø8 + 0.098/+ 0.04; ø8 h8 = ø8 + 0/−0.036)

Fig. 20.51 Tolerancing of a hinge: clearance fit

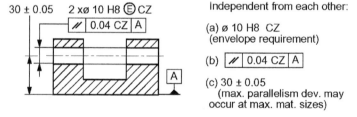

Independent from each other:

(a) ø 10 H8 CZ
(envelope requirement)

(b) [// | 0.04 CZ | A]

(c) 30 ± 0.05
(max. parallelism dev. may
occur at max. mat. sizes)

Fig. 20.52 Tolerances independent of each other

Figure 20.57 shows tolerancing of two coaxial holes to fit with a bolt of ø10 related to the datums A and B.

20.8.2 Coaxiality of bearing surfaces

Figure 20.58 shows tolerancing of a shaft with clearance fits and the relevant gauge.

Figure 20.59 shows tolerancing of the part to fit into the gauge of Fig. 20.58 by using run-out tolerances. The indication of the run-out tolerance excludes the application of

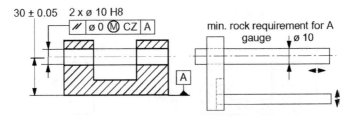

Fig. 20.53 Parallelism tolerance contained in the size tolerance of the holes

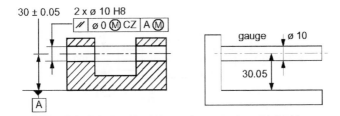

Fig. 20.54 Complex fit: tolerance distributed between hole size and distance

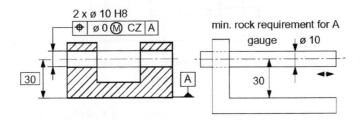

Fig. 20.55 Complex fit: tolerance at the hole size only

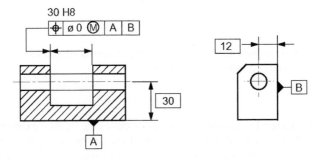

Fig. 20.56 Tolerancing of a hinge gap

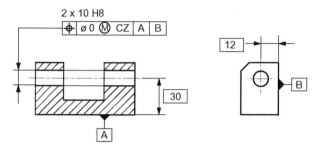

Fig. 20.57 Tolerancing of coaxial holes to fit with a bolt

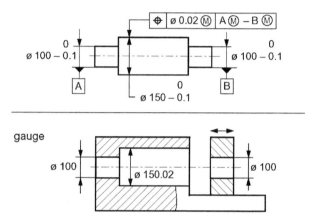

Fig. 20.58 Tolerancing of a shaft with clearance fits and the relevant gauge

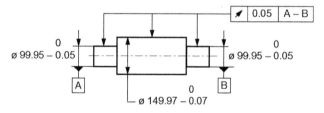

Fig. 20.59 Run-out tolerances of two bearing surfaces related to their common axis and to fit in the gauge of Fig. 20.58

the maximum material requirement. Therefore, to ensure the fit, the maximum material sizes of the three cylindrical features are to be decreased by the value of the run-out tolerance. Note, problems may arise with lobed forms when the part is supported in V-blocks during inspection (see 18.7.4.3 and 18.7.10.1).

20.8.3 Tolerancing of rings, bushes and hubs for fits

Figure 20.60 shows various possibilities for tolerancing of the perpendicularity deviation of a hole in relation to a planar surface.

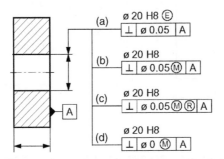

(According to ISO 286: ø20 H8 = ø20 + 0.033/−0)

Fig. 20.60 Perpendicularity tolerances of a hole related to a planar surface

In case (a) the hole must respect the envelope requirement Ⓔ. The actual axis of the hole must be contained in a tolerance cylinder of ø0.05 that is perpendicular to the datum A. The datum A is positioned according to the minimum rock requirement.

In case (b) the hole must respect the maximum material virtual cylinder of diameter given by the maximum material size minus the perpendicularity tolerance (ø20 − 0.05 = ø19.95) that is perpendicular to the datum A. The datum A is positioned according to the minimum rock requirement. However, the actual local sizes of the hole must be between ø20.033 and ø20.

In case (c) the hole and its actual sizes must respect the maximum material virtual cylinder of diameter given by the maximum material size minus the perpendicularity tolerance (ø20 − 0.05 = ø19.95) that is perpendicular to the datum A. The datum A is positioned according to the minimum rock requirement. The actual local sizes of the hole may vary between ø20.033 and ø19.95.

In case (d) the hole must respect the maximum material virtual cylinder of diameter given by the maximum material size minus the perpendicularity tolerance (ø20 − 0 = ø20) that is perpendicular to the datum A. The datum A is positioned according to the minimum rock requirement. The actual local sizes of the hole may vary between ø20.033 and ø20.

Figure 20.61 shows various possibilities for tolerancing of the perpendicularity deviations of a hole in relation to two parallel planar surfaces (hole axis in relation to median plane).

In case (a) the hole must respect the envelope requirement Ⓔ. The actual axis of the hole must be contained in a tolerance cylinder of ø0.05 that is perpendicular to the datum A. The datum A comprises two parallel planes a minimum distance apart enclosing the actual surfaces and orientated according to the minimum rock requirement.

In case (b) the actual hole must respect the maximum material virtual cylinder of diameter given by the maximum material size minus the perpendicularity tolerance (ø20 − 0.05 = ø19.95) that is perpendicular to the datum A. The datum A comprises two parallel planes a minimum distance apart enclosing the actual surfaces and orientated according to the minimum rock requirement.

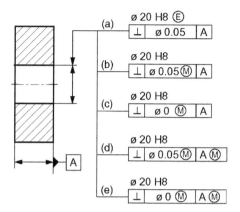

(According to ISO 286: ø20 H8 = ø20 + 0.033/−0)

Fig. 20.61 Perpendicularity tolerance of a hole in relation to two parallel planar surfaces

In case (c) the actual hole must respect the maximum material virtual cylinder of diameter given by the maximum material size minus the perpendicularity tolerance (ø20 − 0 = ø20) that is perpendicular to the datum A. The datum A comprises two parallel planes a minimum distance apart enclosing the actual surfaces and orientated according to the minimum rock requirement.

In case (d) the hole must respect the maximum material virtual cylinder of diameter given by the maximum material size minus the perpendicularity tolerance (ø20 − 0.05 = ø19.95) that is perpendicular to the datum A. Datum A is established by two parallel planes a maximum material size (25) apart. Between these planes, the actual workpiece may float (translate, rotate). In other words, the workpiece must fit into a gauge of maximum material size minus the perpendicularity tolerance at the hole and maximum material size at the width. However, the actual sizes of the hole must be between ø20.033 and ø20.

In case (e) the hole must respect the maximum material virtual cylinder of diameter given by the maximum material size minus the perpendicularity tolerance (ø20 − 0 = ø20) that is perpendicular to the datum A. Datum A is established by two parallel planes a maximum material size (25) apart. Between these planes, the actual workpiece may float (translate, rotate). In other words, the workpiece must fit into a gauge of maximum material size minus the perpendicularity tolerance at the hole and maximum material size at the width (see Fig. 13.4).

Figure 20.62 shows a case similar to Fig. 20.61 (d). However, the actual local sizes of the hole may vary between ø20.033 and ø19.95. The reciprocity requirement is applied to give a recommendation to the manufacturer as to how to distribute the total tolerance on size and geometry.

20.8.4 Rectangular fit

Figure 20.63 shows tolerancing of rectangular features that should fit together.

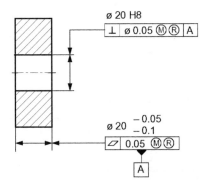

ø 20 H8

ø 20 $^{-0.05}_{-0.1}$

(According to ISO 286: ø20 H8 = ø20 + 0.033/0)

Fig. 20.62 Perpendicularity tolerance of a hole in relation to two parallel planar surfaces: reciprocity requirement applied

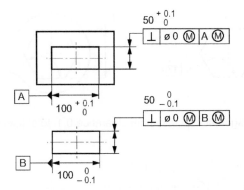

Fig. 20.63 Tolerancing of rectangular features that shall fit together: total tolerance at the size

Without altering the functional requirement shown in Fig. 20.63, it is possible to indicate a manufacturing recommendation as to how the total tolerance on size and geometry can be distributed (reciprocity requirement) (Fig. 20.64; see also 9.3.6).

20.8.5 Hexagon fit

Figure 20.65 shows a hexagon fit with a minimum clearance 0.1. The total tolerance is indicated at the nominal size.

Figure 20.66 shows the same fit. However, the total tolerance is distributed on size and geometry as a manufacturing recommendation (reciprocity requirement) (see also 9.3.6). Sizes and geometry must respect the maximum material virtual condition: 34 for the head and 34.1 for the wrench, which are in the theoretically exact locations (120°) relative to each other.

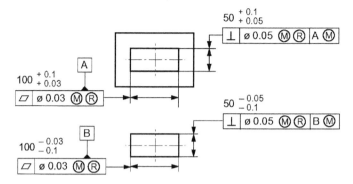

Fig. 20.64 Tolerancing of rectangular features that shall fit together: geometrical tolerance and size tolerance separated as a recommendation

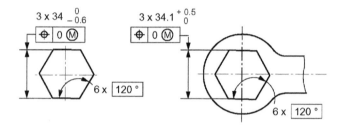

Fig. 20.65 Hexagon fit with minimum clearance 0.1: total tolerance at the size

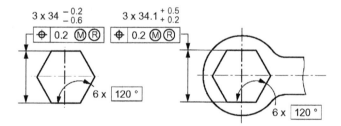

Fig. 20.66 Hexagon fit with minimum clearance 0.1: total tolerance distributed on size and geometry as a recommendation (reciprocity requirement)

Figure 20.67 shows a similar fit using surface profile tolerances for the single plane surfaces.

Figure 20.68 shows the effect of these tolerances. The limitation at the maximum material limit (go gauge) is the same as according to Figs 20.65 and 20.66. At the minimum material limit the geometrical ideal form and orientation apply according to Fig. 20.67, but not according to Figs 20.65 and 20.66. The flatness deviation may occur up to 0.6 or 0.5 respectively (=size tolerance) according to Figs 20.65 and 20.66 and

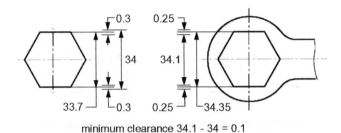

Fig. 20.67 Hexagon fit with minimum clearance 0.1: surface profile tolerances for the single plane surfaces

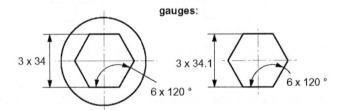

minimum clearance 34.1 - 34 = 0.1

Fig. 20.68 Effect of positional tolerancing according to Fig. 20.67

gauges:

Fig. 20.69 Go gauges for GD&T according to Figs 20.65 and 20.67

0.3 or 0.25 respectively according to Fig. 20.67. According to Fig. 20.67, when the features are at the least material virtual size (34.6) the surfaces must be perfectly planar. This may be an unnecessary restriction and shall be avoided. Tolerancing according to Fig. 20.65 is recommended. See also 23.4.

Figure 20.69 shows the go gauges for GD&T according to Figs 20.65 to 20.67.

20.8.6 Cross fit, only side faces

Figure 20.70 shows a cross fit composed of eight side faces of each part. (The four head faces have larger clearances, and are not included in the fit.) The reciprocity requirement may also be applied (see Fig. 20.66).

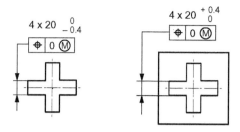

Fig. 20.70 Cross fit: total tolerance at the size

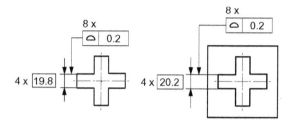

Fig. 20.71 Cross fit: positional tolerances for the single side faces

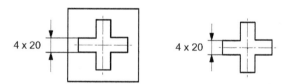

Fig. 20.72 Go gauges for GD&T according to Figs 20.70 and 20.71

Figure 20.71 shows a similar fit using positional tolerances for the single planar surfaces. The limitation at the maximum material limit (go gauge) is the same as according to Fig. 20.70. At the minimum material limit the geometrical ideal form and orientation apply according to Fig. 20.71 but not according to Fig. 20.70.

Figure 20.72 shows the go gauge for GD&T according to Figs 20.70 and 20.71.

20.8.7 Cross fit, all around

Figure 20.73 shows a cross fit covering all faces around. The reciprocity requirement may also be applied (see Fig. 20.64). See also 23.4.

Figure 20.74 shows a similar fit using surface profile tolerances all around for the plane surfaces. This may be an unnecessary restriction (see 20.8.5); tolerancing according to Fig. 20.73 is recommended.

Figure 20.75 shows the go gauges for GD&T according to Figs 20.73 and 20.74.

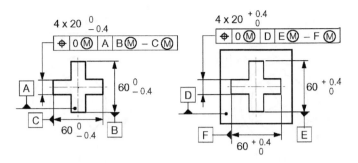

Fig. 20.73 Cross fit: total tolerance at the size

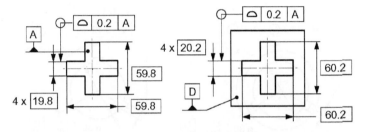

Fig. 20.74 Cross fit: positional tolerances for the single planar surfaces

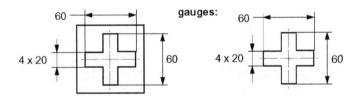

Fig. 20.75 Go gauges for GD&T according to Figs 20.73 and 20.74

20.8.8 Splines

Figure 20.76 shows a spline fit. The total tolerance is indicated at the nominal size. Figure 20.77 shows the same fit, but the total tolerances are distributed on size and geometry as a manufacturing recommendation (reciprocity requirement) (see 9.3.6). Figure 20.78 shows a similar fit using surface profile tolerances for the single planar surfaces. In all three cases the same go gauge applies (gauge sizes 5 and ø30) (Fig. 20.79). Tolerancing according to Fig. 20.78 may be an unnecessary restriction (see 20.8.5); tolerancing according to Fig. 20.76 is recommended. See also 23.4.

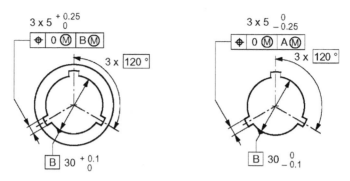

Fig. 20.76 Spline fit: total tolerance at the size

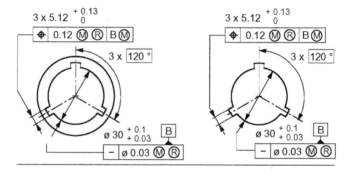

Fig. 20.77 Spline fit: total tolerances distributed on size and geometry (reciprocity requirement)

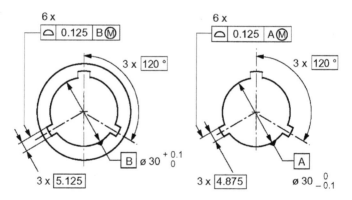

Fig. 20.78 Spline fit: positional tolerances for the single planar surfaces

20.8.9 Plug and socket fit

Figure 20.80 shows a plug and socket fit. The total tolerance is indicated at the nominal size of the cylindrical feature.

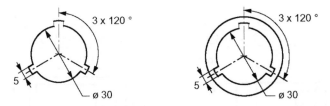

Fig. 20.79 Go gauges for D&T according to Figs 20.76 to 20.78

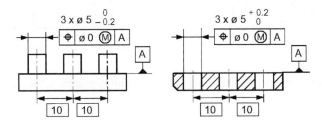

Fig. 20.80 Plug and socket fit: total tolerance at the size

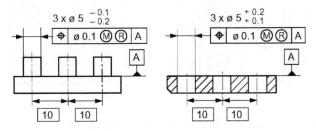

Fig. 20.81 Plug and socket fit: total tolerance distributed on size and position (reciprocity requirement)

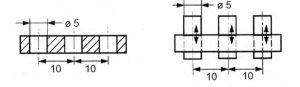

Fig. 20.82 Go gauge for D&T according to Figs 20.80 and 20.81

Figure 20.81 shows the same fit, but the total tolerance 0.2 is distributed on size and position as a manufacturing recommendation (reciprocity requirement) (see 9.3.6).

Figure 20.82 shows the go gauges for GD&T according to Figs 20.80 and 20.81. Because the positional tolerance is related to the primary datum A the workpiece surface

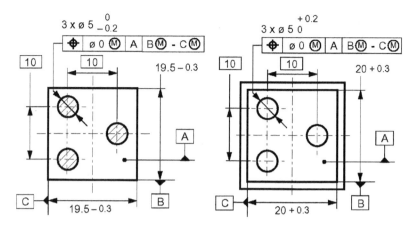

Fig. 20.83 Plug and socket fit with frame fit: total tolerance at the sizes

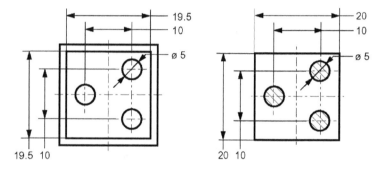

Fig. 20.84 Go gauges for GD&T according to Fig. 20.83

A must be orientated according to the minimum rock requirement relative to the gauge surface A.

Figures 20.83 and 20.84 show a plug and socket fit where the socket has a frame and forms a further fit with the plug circumference, which is related to the fits of the pins and holes. At the frame is a minimum clearance of 0.5 that is the difference between the maximum material sizes. See also 23.4.

Figure 20.84 shows the go gauges for GD&T according to Fig. 20.83.

Figure 20.85 shows tolerancing of a socket similar to that shown in Fig. 20.80, but with additional positional tolerances related to datums A, B and C. The holes must respect the maximum material virtual cylinder of ø20 (perpendicular to datum A and 40 apart) and the maximum material virtual cylinder of ø19.5 which are in the theoretical exact orientation and location in relation to the datums A, B and C. See also 23.4 for the 0 Ⓜ tolerance.

Figure 20.86 shows the go gauges for GD&T according Fig. 20.85. The workpiece surface A must be orientated according to the minimum rock requirement relative to the surface A of gauge 1a; then gauge 1b must fit in the workpiece. Next the workpiece must be positioned according to the minimum rock requirement relative to surfaces A

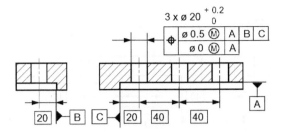

Fig. 20.85 Tolerancing of a socket (similar to Fig. 20.80) relative to a datum system: tolerances for the position of the holes in relation to each other within the size tolerance

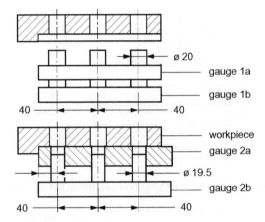

Fig. 20.86 Separated go gauges (go gauge set) for D&T according to Fig. 20.85

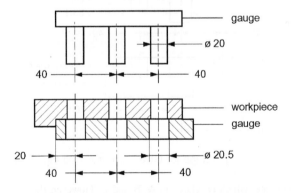

Fig. 20.87 Combined go gauge for GD&T according to Fig. 20.85

and B and relative to surface C of gauge 2a (see 3.4); then gauge 2b must fit in the workpiece.

Both gauging procedures may be combined. Figure 20.87 shows the appropriate gauge. The holes in the gauge are larger than the maximum material size by the amount of the

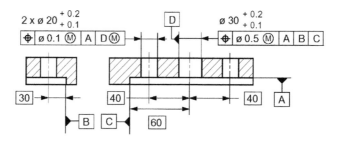

Fig. 20.88 Tolerancing of a socket relative to a datum (guiding) hole A that is toleranced relative to a datum system

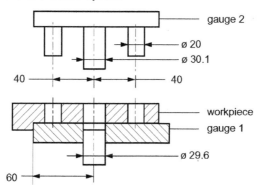

Fig. 20.89 Separated go gauges (go gauge set) for GD&T according to Fig. 20.88

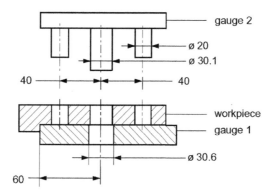

Fig. 20.90 Combined go gauge for GD&T according to Fig. 20.88

positional tolerance, which is related to A, B and C. However, this gauging does not inspect the perpendicularity requirement of 0 Ⓜ related to A.

Figure 20.88 shows positional tolerancing of two holes related to a datum hole D. Datum hole D is related to a datum system A, B and C. See also 23.4 for the 0.1 Ⓜ tolerance.

Figure 20.89 shows the appropriate go gauge set.

Both gauging procedures may be combined. Figure 20.90 shows the appropriate gauge. The gauge plug diameter for the datum hole D is equal to the maximum material

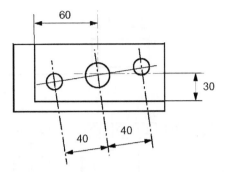

Fig. 20.91 Permissible orientation of the holes according to Fig. 20.88

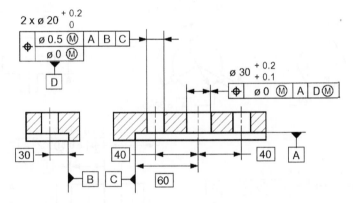

Fig. 20.92 Two holes positional toleranced in relation to the datum system A, B and C in order to limit the effect of Fig. 20.91

size. The hole in gauge 1 has maximum material size plus positional tolerance. The workpiece must be positioned according to the minimum rock requirement relative to the surfaces A, B and C of gauge 1 (see 3.4). However, this gauging does not inspect the perpendicularity requirement 0.1 Ⓜ related to A.

Tolerancing according to Fig. 20.88 does not limit the inclination of the hole pattern in relation to the datums B and C, Fig. 20.91.

In order to limit the effect shown in Fig. 20.91, at least two holes must be positional toleranced in relation to the datum system A, B and C (Fig. 20.92).

20.8.10 Synopsis tolerancing of clearance fits

Figure 20.93 shows a synopsis of frequent cases of clearance fits toleranced appropriately for the function.

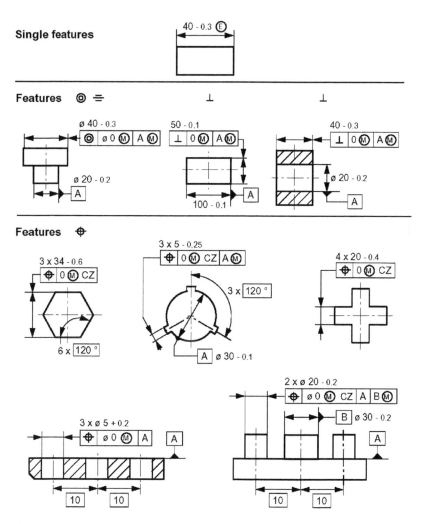

Fig. 20.93 Clearance fits: functional cases, Ⓔ, ◎, ≟, ⊥, ⊕

Figure 20.94 shows a synopsis of frequent functional cases of coaxial holes with clearance fits. The relevant gauges demonstrate the function. In the upper example the indications (a) and (b) have the same meaning.

20.9 Interference fits and kinematics

20.9.1 Coaxial holes

Figure 20.95 shows a hinged frame with holes ø8 N 9 (0/−0.036) Ⓔ that shall fit together with a bolt ø8 h 9 (0/−0036) Ⓔ (transition fit).

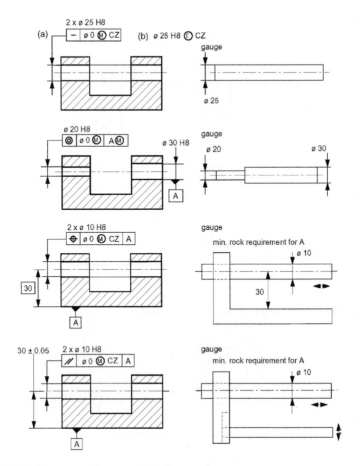

Fig. 20.94 Clearance fits: coaxial holes

Figure 20.95 (bottom) shows the extreme case when all actual sizes are at the worst limit. In this case the coaxiality deviation (eccentricity) of the two holes is 0.01. The bolt is guided without clearance by the right hole, and must deflect when entering the left hole. Here the indication of the maximum material requirement Ⓜ would be detrimental because it would result in a greater deflection of the bolt necessary for the fit. The maximum material requirement would allow larger coaxiality deviations with larger holes, although the bolt may already be guided in one hole without clearance.

Inspection can be performed with two mandrels that fit into the holes without clearance. Compare with Fig. 20.51.

20.9.2 Geometrical ideal form at least material size

Figure 20.96 shows tolerancing when the geometrical ideal form (here a cylinder) at least material size (here the maximum size) must be respected. The surface must not encroach upon this boundary in the direction to the material.

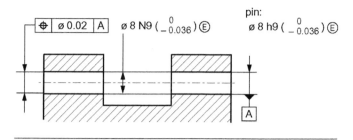

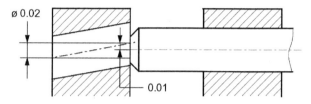

Fig. 20.95 Tolerancing of a hinge: transition fit

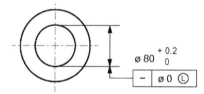

Fig. 20.96 Hole with minimum material requirement

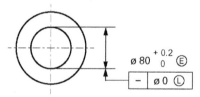

Fig. 20.97 Hole with envelope requirement and least material requirement

Figure 20.97 shows tolerancing when, in addition to the geometrical ideal boundary at least material size, the envelope requirement Ⓔ shall apply (i.e. when the surface must be contained between the geometrical ideal boundaries at maximum material size and at least material size, e.g. for interference fits).

Figure 20.98 has a similar meaning as Fig. 20.97. In Fig. 20.98 the boundaries are coaxial, while in Fig. 20.97 not necessarily so.

20.9.3 Gears

Figure 20.99 shows tolerancing of a gear. Because of a clearance fit, the maximum material requirement is applied to the hole and to the endfaces (width). Because of

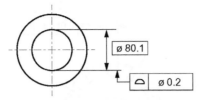

Fig. 20.98 Hole with surface profile tolerance; geometrical ideal forms with maximum and least material sizes must be respected

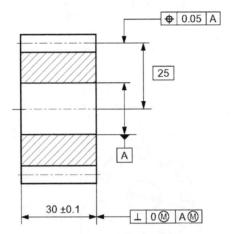

Fig. 20.99 Tolerancing of a gear

kinematic considerations at the tooth mesh contact, the maximum material requirement applied to the pitch diameter would be detrimental, resulting in greater offset of the tooth contact points.

20.10 Distances and thicknesses

20.10.1 Spacings

According to Fig. 20.100, the median faces of the slots shall not deviate by more than ±0.025 from the symmetrical position in relation to the axis of the datum hole B and the theoretical exact spacing. The positional tolerance is independent from the actual sizes of the slots and of the hole.

According to Fig. 20.101, slots and hole must fit into a geometrical ideal counterpart (go gauge) of maximum material size at the hole and maximum material size (3.35) at the slots. With this requirement the positional tolerance depends on the actual sizes and actual forms of the slots and of the hole. The actual size of the slot may be 3.35 when the slot is in the geometrical ideal form and position. See also 23.4.

According to Fig. 20.102, the position of the right faces of the slots in relation to the axis of the datum hole B is closely toleranced (0.02). The widths of the slots are less so

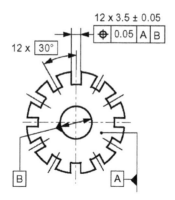

Fig. 20.100 Spacing: positional toleranced slot median face

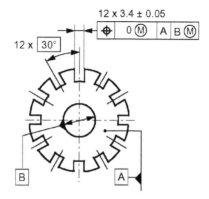

Fig. 20.101 Spacing: positional toleranced slot median face with maximum material requirement

($\pm$0.05). The surface profile tolerances of the right slot faces are independent of the actual sizes of the slots and of the hole.

According to Fig. 20.103, the slot faces must respect the geometrical ideal least material virtual condition (boundary 4.05) whose median plane contains the axis of the datum hole B. "Respect" here means that the boundary lies entirely within the material. The positional tolerance is 0.5 at least material size (3.55), and becomes larger with smaller slots. The positional tolerance is independent of the actual size of the hole.

20.10.2 Hole edge distance

Figure 20.104 shows (left) tolerancing of distances by linear dimensional tolerances and (right) a possible interpretation. The shortest distances from the hole axes to the side face are toleranced. There are also other possible interpretations. Therefore this type of tolerancing is not recommended for precision parts.

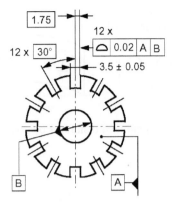

Fig. 20.102 Spacing: surface profile toleranced slot face

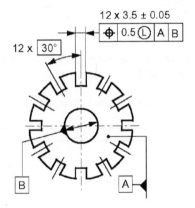

Fig. 20.103 Spacing: positional toleranced slot median face with least material requirement

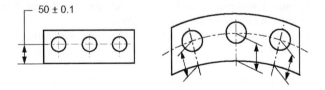

Fig. 20.104 Tolerancing of hole face distances by linear tolerances

Figure 20.105 shows (left) tolerancing of distances by positional tolerances related to a datum face A and (right) the explanation. The distances from the hole axes to the datum, i.e. practically to a contacting gauge plate (simulated datum features), are toleranced.

20.10.3 Equal thickness

Some functions (e.g. of thin discs of multiple-disc clutches) require small limits on the variation of the actual local sizes (e.g. of thickness) within the same feature (e.g. two

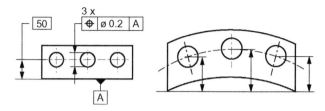

Fig. 20.105 Tolerancing of hole face distances by positional tolerances related to a datum plane

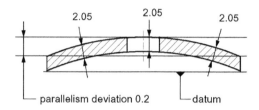

Fig. 20.106 Workpiece with considerable parallelism deviation but almost no variation in its actual local sizes

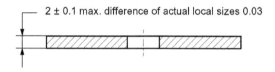

Fig. 20.107 Tolerancing of equal thickness

parallel planar faces) of a single workpiece. The size deviations of different workpieces, however, may vary over a larger range. The parallelism tolerance is not appropriate for this purpose, since a workpiece may have a considerable parallelism deviation but almost no variation in its actual local sizes (Fig. 20.106).

A symbol for equal thickness is not yet standardized internationally. Therefore the following indication may be used "max. permissible difference of actual local sizes..." (Fig. 20.107).

At present it is under discussion whether to standardize the drawing indication for equal thickness as, for example, $2 \pm 0.1 \, \Delta \, 0.03$.

20.10.4 Minimum wall thickness

When the function requires a minimum wall thickness, and the wall thickness and position of the wall depend on each other, then the permissible positional deviation of the wall becomes larger the more the wall deviates from its least material size (the more the wall becomes thicker).

Figure 20.108 shows tolerancing to ensure a minimum wall thickness (2.705) between hole and recess. When the hole is at its least material size (ø4.2), the hole axis must be

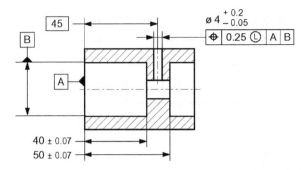

Fig. 20.108 Positional tolerance with least material requirement to ensure a minimum wall thickness

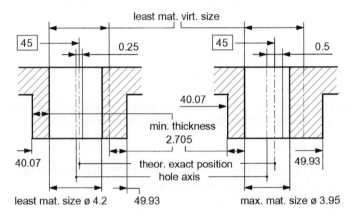

Fig. 20.109 Calculation of the minimum wall thickness 2.705 for tolerancing according to Fig. 20.108: left, hole least material size; right, maximum material size

contained in a tolerance zone ø0.25. In the worst case, when the hole axis has the largest positional deviation, the remaining wall thickness is 2.705 (Fig. 20.109 left). When the hole is at its maximum material size (ø3.95), the tolerance zone of the hole axis increases by the amount of the size tolerance (ø0.25) to become ø0.5. Then also in the worst case the remaining wall thickness is 2.705 (Fig. 20.109 right).

In other words, the hole must respect (it must be contained within the material) the least material virtual cylinder of diameter given by the least material size plus the positional tolerance (4.2 + 0.25 = 4.45) that is in the theoretical exact position. Between the least material virtual cylinder and the recesses there remains a minimum wall thickness of 2.705 (Fig. 20.109).

Figure 20.110 shows tolerancing to ensure a minimum wall thickness of 8.5. The permissible positional deviation of the inner cylinder increases when the diameter of the inner cylinder decreases. The permissible positional deviation of the outer cylinder increases when the diameter of the outer cylinder increases.

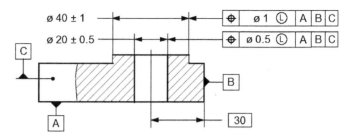

Fig. 20.110 Positional tolerances with least material requirements to ensure a minimum wall thickness

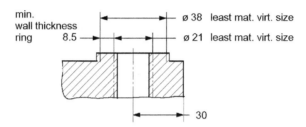

Fig. 20.111 Calculation of the minimum wall thickness 2.5 for tolerancing according to Fig. 20.110: left, features at least material size; right, at maximum material size

Figure 20.111 shows the theoretical exact position of the least material virtual cylinders that must be respected by the workpiece surfaces.

If appropriate, the reciprocity requirement Ⓡ after Ⓛ may be applied additionally (see 11.4).

20.11 Geometrical ideal form at maximum and least material sizes

In Fig. 20.112(a) the toleranced cylindrical feature should not violate the maximum material virtual condition (limiting boundary) of ø20. In order to meet this requirement, the manufacturer must distribute the total tolerance of 0.3 on size and form (e.g. 0.2 for size deviations and 0.1 for form deviations). In Fig. 20.112 (b) this distribution is already indicated. However, the indications of Ⓜ and Ⓡ after the straightness tolerance indicate that the straightness tolerance and the size tolerance are increased to the extent allowed by the maximum material virtual condition.

Both indications, Fig. 20.112(a) and (b), specify the same functional requirement (maximum material virtual condition of ø20 and actual local sizes within ø20 0/−0.3). Figure 20.112(b) gives further information to the manufacturer on the recommended distribution of the total tolerance on size and form, i.e. it provides a symbolization for communication between production planning and workshop.

Figure 20.113(a) and (b) gives analogue information on the least material side. The cylindrical feature shall not violate the least material virtual condition (limiting boundary) of ø19.7. In order to meet this requirement, the manufacturer must distribute the total tolerance of 0.3 on size and form (e.g. 0.2 for size deviations and 0.1 for form deviations).

In Fig. 20.113(b) this distribution is already indicated. However, the indications of Ⓛ and Ⓡ after the straightness tolerance indicate that the straightness tolerance and the size tolerance are increased to the extent given by the least material virtual condition.

Figure 20.114(a) and (b) show the drawing indications when both limiting boundaries (maximum material virtual condition and least material virtual condition) are to be specified. In order to meet these requirements, the manufacturer must distribute the total tolerance of 0.3 on size and form (e.g. 0.1 for size deviations and 0.1 on each side (maximum material side and least material side) for form deviations). In Fig. 20.114 (b) this distribution is already indicated. However, the indications Ⓡ after the symbols Ⓜ and Ⓛ after the straightness tolerance indicate that the straightness tolerances and the size tolerance are increased to the extent given by the maximum material virtual condition and the least material virtual condition.

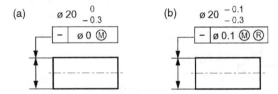

Fig. 20.112 Maximum material requirement: (a) without and (b) with reciprocity requirement

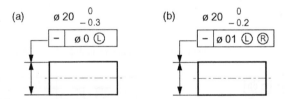

Fig. 20.113 Least material requirement: (a) without and (b) with reciprocity requirement

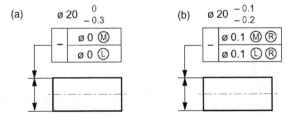

Fig. 20.114 Maximum material requirement and least material requirement simultaneously (a) without and (b) with reciprocity requirement

In Figs 20.112 to 20.114 the left indications (a) specify the same functional requirements as the right indications (b); for example, Fig. 20.112(a) and (b) specify the
maximum material virtual condition of ø20 and actual local sizes to be within ø20 $0/-0.3$.
However, the right indications (b) give further information to the manufacturer on the recommended (not mandatory) distribution of the total tolerance on size and form.

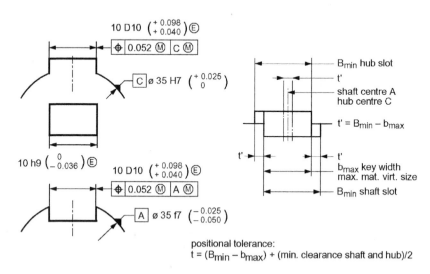

positional tolerance:
$t = (B_{min} - b_{max}) +$ (min. clearance shaft and hub)/2

Fig. 20.115 Key assembly with only one floating key

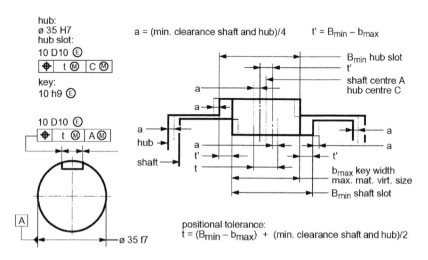

positional tolerance:
$t = (B_{min} - b_{max}) +$ (min. clearance shaft and hub)/2

Fig. 20.116 Additional positional tolerance $2a$ = (minimum clearance between
shaft and hub)/2 shown at the example of a floating key

The same functional requirements (ø0 Ⓜ and ø0 Ⓛ) can be expressed by using the theoretical exact dimension 19.85 and a surface profile tolerance 0.15. But then there is no possibility for the use of Ⓡ in order to split the tolerance in permissible deviations of size and form.

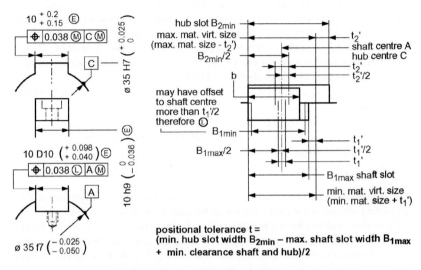

Fig. 20.117 Key assembly with clearance fits and only one fixed key

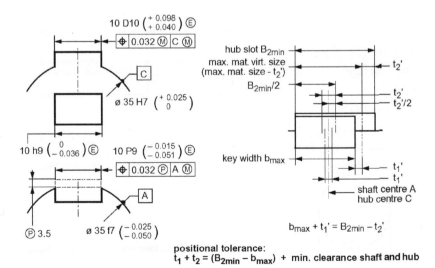

Fig. 20.118 Key assembly with interference/clearance fits, only one key

20.12 Keyways

Keyways may be assembled with keys using clearance fits with floating keys (Fig. 20.115) or clearance fits with fixed keys (Fig. 20.117) or interference fits in the shaft and clearance fits in the hub (Fig. 20.118). The right parts of these figures show the virtual conditions resulting from the indications given in the left parts of the figures. Usually the positional tolerances of the slots in shaft and hub are chosen $t_1 = t_2 = (t_1 + t_2)/2 = t$.

When there is only one key and a minimum clearance between shaft and hub (clearance fit) each slot can have an additional eccentricity component of 1/4 of the minimum clearance between shaft and hub or a positional tolerance component of 1/2 of the minimum clearance between shaft and hub (Fig. 20.116). Then the positional tolerance is $t = t' +$ (minimum clearance shaft and hub)/2.

When there are more than one key $t = t'$, $t_1 = t'_1$, $t_2 = t'_2$ applies, i.e. the positional tolerance is

$$t = t' = (B_{min} - b_{max}) \qquad \text{in Fig. 20.115} \quad \text{see Fig. 20.119}$$
$$t = t_1/2 + t_2/2 = (B_{2min} - B_{1max})/2 \quad \text{in Fig. 20.117} \quad \text{see Fig. 20.120}$$
$$t = t_1/2 + t_2/2 = (B_{2min} - b_{max})/2 \quad \text{in Fig. 20.118} \quad \text{see Fig. 20.121}$$

Then the clearance between shaft and hub allows additional positional deviation of the slot pattern relative to the datum but not of the slots relative to each other.

When shaft and hub have a transition fit or an interference fit, e.g. m6/h6, there is no positional tolerance component resulting from the deviation from the maximum material limits of size of the shaft and hub diameters and no ⓂＭ at the datums. In these cases $t = t'$, $t_1 = t'_1$, $t_2 = t'_2$ apply, i.e. the formulae given above.

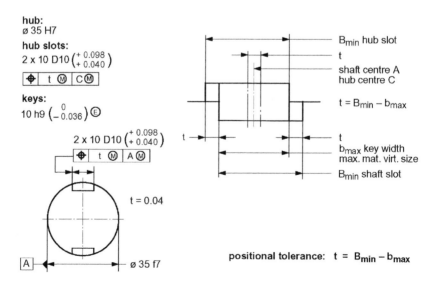

Fig. 20.119 Key assembly with two or more floating keys

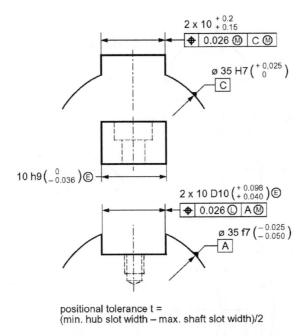

positional tolerance t =
(min. hub slot width – max. shaft slot width)/2

Fig. 20.120 Key assembly with two or more fixed keys with clearance fits

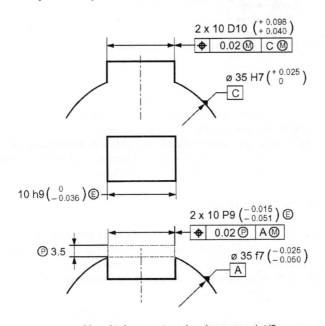

positional tolerance t = min. clearance slot/2

Fig. 20.121 Key assembly with two or more keys with interference/clearance fits

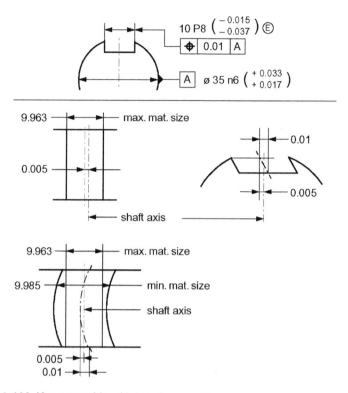

Fig. 20.122 Key assembly with interference fits

When the minimum clearance = 0 between keyway and key is acceptable the 0 Ⓜ method, similar to that shown in Fig. 20.76, may be used.

Figure 20.122 shows an example of a key assembly with interference fits at the key and at the shaft and hub. Here all geometrical deviations lead to deformations of the parts. Figure 20.122 shows the extreme possible deviations.

20.13 Holes for fasteners

Fastener assemblies may be designed with through holes and (floating) bolts and nuts (Fig. 20.123) or with threaded holes and head screws (Fig. 20.124) or with threaded holes and (fixed) stud screws (Fig. 20.125). In the cases of Figs 20.124 and 20.125 the through holes in the other flange have the same tolerance but without Ⓟ.

20.14 Cones and wedges

Figure 20.126 shows the tolerancing of a cone when form, orientation (radial) and location (axial) deviations are limited by the profile tolerance (0.05) of the cone.

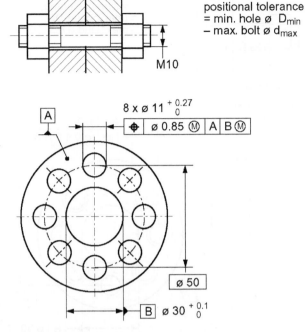

positional tolerance
= min. hole ø D_{min}
– max. bolt ø d_{max}

M10

$8 \times ø\ 11\ ^{+0.27}_{0}$

| ⊕ | ø 0.85 Ⓜ | A | B Ⓜ |

A

ø 50

B | ø 30 $^{+0.1}_{0}$

Fig. 20.123 Through holes for floating fasteners (bolts and nuts)

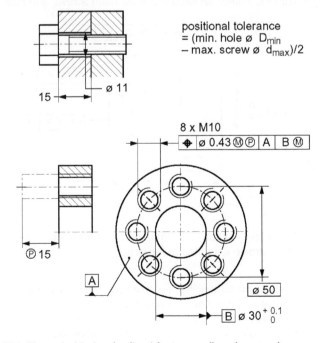

positional tolerance
= (min. hole ø D_{min}
– max. screw ø d_{max})/2

ø 11

15

$8 \times$ M10

| ⊕ | ø 0.43 ⓂⓅ | A | B Ⓜ |

Ⓟ 15

A

ø 50

B | ø 30 $^{+0.1}_{0}$

Fig. 20.124 Threaded holes for fixed fasteners (head screws)

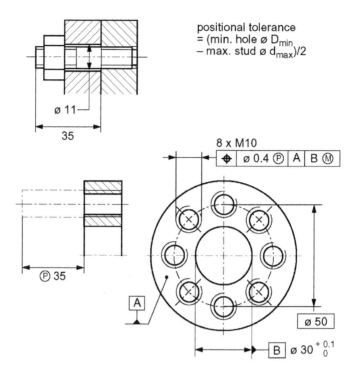

positional tolerance
= (min. hole ø D_{min}
– max. stud ø d_{max})/2

Fig. 20.125 Threaded holes for fixed fasteners (stud screws and nuts)

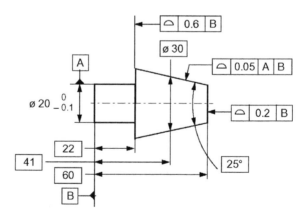

Fig. 20.126 Tolerancing of cones, orientation and location deviations are limited by the surface profile tolerance of the cone

Figure 20.127 shows the tolerancing of a cone when form, orientation and location deviations of the cone are limited by separate tolerances.

Figure 20.128 shows tolerancing according to Fig. 20.126 when in addition auxiliary diameters and their tolerances for manufacturing (which are easy to measure

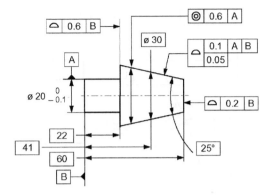

Fig. 20.127 Tolerancing of cones: form, orientation and location deviations are limited by separate tolerances

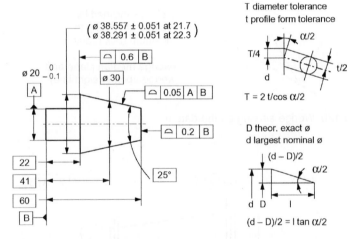

Fig. 20.128 Tolerancing of a cone with auxiliary dimensions for manufacturing purposes

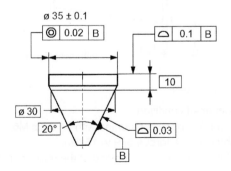

Fig. 20.129 Cone serving as the datum

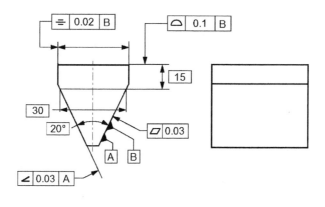

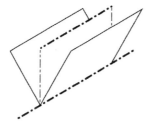

datum B composed of:

2 planes inclined by 20°
located according to
minimum rock requirement

(wedge median plane and
wedge apex straight line)

Fig. 20.130 Wedge serving as the datum

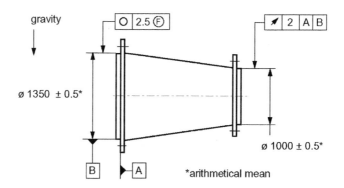

ISO 10 579-NR

Restrained condition:
datum A surface is mounted with 64 bolts M6
tightened to a torque of 9 to 15 Nm;
datum B surface is restrained at the max. mat. limit

Fig. 20.131 Tolerancing of a flexible part

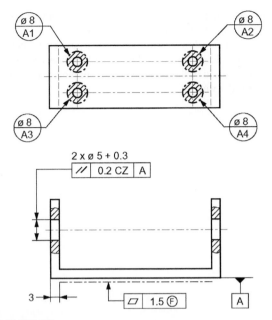

ISO 10 579-NR
Restrained condition: A1 to A4 coplanar

Fig. 20.132 Flexible part with more than three datum targets for the plane surface

during manufacturing) in parentheses are indicated. The limit dimensions are dependent on the distance from the datum B. The figure shows the formulae for the calculation.

Figure 20.129 shows a cone serving as the datum (B) for tolerancing the coaxiality deviation of the short cylinder (too short to serve as the datum).

Figure 20.130 shows a wedge serving as the datum (B) for tolerancing the symmetry deviation of a short prism (too short to serve as the datum).

See also 5.

20.15 Flexible parts

Figure 20.131 shows an example of tolerancing of a flexible part. In the free state, subject only to gravity in the indicated direction, the cylinder B must meet the roundness tolerance of 2.5. Its averaged diameter must be within 1350 ± 0.5 (calculated from at least four equally spaced local sizes or calculated from the circumference). The right cylinder must meet the run-out tolerance of 2 when restrained as specified. Its averaged diameter must be within 1000 ± 0.5 (The average diameter does not depend on whether the free state or restrained condition applies).

For flexible parts, depending on the function, it is possible to specify more datum targets than the number described in 3.4. Figure 20.132 shows an example where for the plane surface four instead of three datum targets are indicated because the part will be assembled in a similar way.

21

Differences between ISO Standards and other Standards

21.1 ASME Y14.5M – 1994, ANSI B89.3.1

In the following the terminology of ASME Y14.5 is used (see Table 21.3).

21.1.1 Symbols

The National Standard of the United States of America ASME Y14.5M–1994 specifies in addition to or deviating from ISO 1101 the symbols and drawing indications shown in Tables 21.1 and 21.2.

Table 21.1 Symbols according to ASME Y14.5M

Symbol	Designation	Interpretation
⌒	ALL AROUND	Profile all around, applicable to the bounded line shown
ALL OVER ⌓	ALL OVER	Everywhere, applicable to all surfaces (e.g. of a casting)* (ASME Y14.8)
– A –	DATUM	Datum (former ANSI Y14.5) acc. to ASME Y14.5 as ISO 1101: ▶ A
Ⓢ	REGARDLESS OF FEATURE SIZE	According to ANSI Y14.5 - 1982 to be indicated with positional tolerances, if applicable According to ASME Y14.5 the meaning of Ⓢ applies without indication (as according to ISO 1101)
AVG	AVERAGE	Arithmetical mean (e.g. of the actual sizes of a flexible ring, see flexible parts)
Ⓣ	TANGENT	Tolerance applies to the contacting (tangential) element
◄►	BETWEEN	Tolerance applies to a limited segment of a surface between designated extremities
CR	CONTROLLED RADIUS	Controlled radius ("fair curve without reversals, radii at all points within tolerance")
⟨ST⟩	STATISTICAL TOLERANCING	Statistical tolerance SPC required

*Not specified in ASME Y14.5M but usual in US practice.

The symbol "CZ" is not standardized according to ASME Y14.5. When it is used in drawings referenced to ASME Y14.5 the reference to ISO 1101 should also be introduced into the drawing title block, together with the remark: In cases of contradiction ASME Y14.5 applies.

21.1.2 Rules

ASME Y14.5M deviates in some rules from the specifications given in the ISO Standards.
ASME Y14.5M specifies the following rules:

Rule #1: Where only a tolerance of size is specified, the limits of size of an individual feature prescribe the extent to which variations in its geometrical form, as well as size, are allowed.

In other words, for individual features of size the envelope requirement applies without drawing indication. Exceptions are as follows:

- stock materials (such as bars, sheets, tubing, structural shapes) for which straightness, flatness or other geometrical tolerances are standardized and which remain in the "as-furnished" condition on the finished part;

Table 21.2 Drawing indications according to ASME Y14.5M

Drawing indication	Designation	Interpretation
	UNILATERAL TOLERANCE INSIDE	
	UNILATERAL TOLERANCE OUTSIDE	
	BILATERAL TOLERANCE UNEQUAL DISTRIBUTION	
2 SURFACES	ASME Y14.5: ISO 1101:	
2 x	ASME Y14.5: ISO 1101: COPLANARITY	

- (non-rigid) parts subject to free state variation in the unrestrained condition (see 12);
- features with a straightness tolerance to the axis or a flatness tolerance to the median face.

Features with straightness tolerances to the generator lines are not exceptions. The envelope requirement applies as well as the straightness tolerance. Both requirements have to be respected.

Rule #1 applies only to individual features of size. The rule does not apply to deviations of orientation or location or run-out of one feature related to another feature.

Rule #1 is similar to the drawing title box indication ISO 2768 ... E and to the German Standard DIN 7167. The difference is that ASME14.5M rule #1 excludes also those features with a straightness tolerance of the axis indicated that is smaller than the size tolerance (the other standards do not exclude those features).

Rule for radii: A radius symbol R (preceding the dimension) creates a zone defined by two arcs (the minimum and maximum radii) that are tangent to the adjacent surfaces. The part surface must lie within this zone (Fig. 21.1).

"A controlled radius symbol CR (preceding the dimension) creates a tolerance zone defined by two arcs (the minimum and maximum radii) that are tangent to the adjacent surfaces. When specifying a controlled radius, the part contour within the crescent-shaped tolerance zone must be a fair curve without reversals. Additionally, radii taken at all points on the part contour shall neither be smaller than the specified minimum limit nor larger than the maximum limit (see Fig. 21.2)."

There is not yet a precise definition of the radius tolerance in the ISO standards.[*]

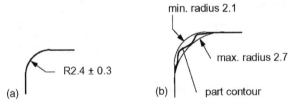

Fig. 21.1 Radius tolerance: (a) drawing indication, (b) interpretation

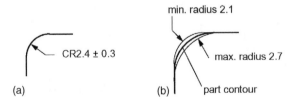

Fig. 21.2 Controlled radius tolerance CR: (a) drawing indication, (b) interpretation

Rule for angular surfaces: "Where an angular surface is defined by a combination of a linear dimension and an angle, the surface must lie within a tolerance zone represented by two non-parallel planes [see Fig. 21.3]. The tolerance zone will become wider as the distance from the apex of the angle increases. Where a tolerance zone with parallel

[*] A Technical report ISO TR 16 570 is in preparation.

boundaries is desired, a basic angle may be specified as in Fig. 21.4. Additionally, an angularity tolerance may be specified within these boundaries (Fig. 21.5)."

In Figs 21.3 and 21.4 it is indicated by the symbol that the dimension between the two features shall originate from the lower feature. The high points of the surface indicated as the origin define a plane of measurement.

Only tolerancing according to Fig. 21.5 complies with international practice, and is therefore recommended. Tolerancing according to Fig. 21.4 is not in accordance with ISO 1101. Drawing indications according to Fig. 21.3 have a different meaning according to ISO 8015 (see Fig. 3.45).

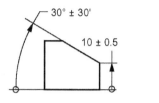 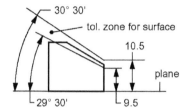

Fig. 21.3 Tolerancing of an angular surface using a combination of linear and angular dimensions, according to ASME Y14.5

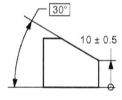 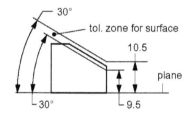

Fig. 21.4 Tolerancing of an angular surface with a basic angle

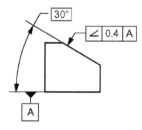 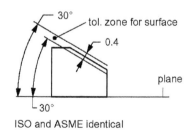

ISO and ASME identical

Fig. 21.5 Tolerancing of an angular surface with an angularity tolerance

Former rule #2 ANSI Y14.5M – 1982 (positional tolerance rule): With tolerances of position "regardless of feature size" Ⓢ, the maximum material requirement Ⓜ or the least material requirement Ⓛ must be specified on the drawing with respect to the individual tolerance, datum reference, or both, as applicable.

According to a former rule (USASI Y14.5 – 1966 and ANSI Y14.5 – 1973), with positional tolerances, the maximum material requirement Ⓜ was always applied to the toleranced feature as well as to the datum feature, without any indication of Ⓜ. This was not in line with the international practice, according to which Ⓢ always applies without the indication Ⓢ, if not otherwise specified (with Ⓜ or Ⓛ), ISO 8015 and ISO 2692.

Former rule #3 ANSI Y14.5M – 1982 (regardless of feature size rule for all other geometrical tolerances than positional tolerances): For all geometrical tolerances, other than positional tolerances, "regardless of feature size" Ⓢ applies, with respect to the individual tolerance, datum reference, or both, where no modifying symbol is specified.

Without the exception this coincides with the International Standards ISO 8015, ISO 2692 and ISO 1101. This rule was later covered by the new rule #2 ASME Y14.5M – 1994.

New rule #2 ASME Y14.5M – 1994: Regardless of feature size applies, with respect to the individual tolerance, datum reference, or both, where no modifying symbol is specified. Ⓜ or Ⓛ must be specified on the drawing where it is required.

This rule #2 coincides with ISO practices (ISO 1101, ISO 2692 and ISO 8015).

New rule #2a ASME Y14.5M – 1994 (alternative praxis): For a tolerance of position Ⓢ **may** be specified on the drawing with respect to the individual tolerance, datum reference, or both as applicable.

This new rule #2a is not in line with international practice. The symbol Ⓢ does not exist in the ISO standards. Regardless of feature size, applies everywhere, if not otherwise specified (with Ⓜ or Ⓛ).

Pitch diameter rule (rule #4): Each tolerance of orientation or position (location) and datum reference specified for a screw thread applies to the axis of the thread derived from the pitch cylinder. Where an exception to this practice is necessary, the specific feature of the screw thread (such as MINOR DIA or MAJOR DIA) shall be stated beneath the feature control frame, or beneath or adjacent to the datum feature symbol, as applicable.

Each tolerance of orientation or location and datum reference specified for gears and splines must designate the specific feature of the gear or spline to which each applies (such as MAJOR DIA, PITCH DIA or MINOR DIA). This information is stated beneath the feature control frame or beneath the datum feature symbol, as applicable.

The new edition of ISO 1101-2004 gives a similar rule but different indications (MD for major, LD for minor (least) and PD for pitch diameter).

Datum virtual condition rule: Where a datum feature of size is applied on an MMC basis, machine and gauging elements in the processing equipment, which remain constant in size, may be used to simulate a true geometrical counterpart of the feature and to establish the datum. In each case, the size of the true geometric counterpart (simulated datum) is determined by the specified MMC limit of size of the datum feature or its MMC virtual condition, where applicable.

Where a primary or single datum feature of size is controlled by a roundness or cylindricity tolerance, the size of the true geometric counterpart used to establish the simulated datum is the MMC limit of size. Where a straightness tolerance is applied on an MMC basis, the size of the true geometric counterpart used to establish the simulated datum is the MMC virtual condition of the datum feature. Where a straightness toler-ance is applied on an RFS basis, the size of the true geometric counterpart used to establish the simulated datum is the applicable inner or outer boundary (see 21.1.10).

Where secondary or tertiary datum features of size in the same datum reference frame are controlled by a specified tolerance of location or orientation with respect to each other, the size of the true geometric counterpart used to establish simulated datum is the virtual condition of the datum feature.

Analysis of tolerance controls applied to a datum feature is necessary in determining the size for simulating its true geometrical counterpart. Consideration must be given to the effects of the difference in size between the applicable virtual condition of a datum feature and its MMC limit of size. Where a virtual condition equal to MMC is the design requirements a zero geometrical tolerance at MMC is to be specified (0 Ⓜ).

In other words, for a datum feature of size used as a datum and for a feature of size with a geometrical tolerance applied to its axis or median face, a virtual condition (maximum material virtual condition) applies. The size of the virtual condition depends on the maximum material size and on the value of the geometrical tolerance applied to the axis or median face of the datum. When the datum is used as a secondary or terti-ary datum, the orientation and location of the virtual condition depend on the primary and secondary datums.

For datum features of size (features with an axis or median plane) with the indica-tion Ⓜ behind the datum letter in the tolerance frame, the following applies:

(a) When for the datum feature with an axis a form tolerance to the feature (not to its axis) is indicated (e.g. roundness tolerance, straightness tolerance of the generator line), the maximum material virtual size is equal to the maximum material size (Fig. 21.6).

 When for a datum feature with an axis, a form tolerance to the axis is indicated (e.g. straightness tolerance of the axis), the maximum material virtual size is equal to the maximum material size plus (shafts) or minus (holes) the form tolerance of the axis (Figs 21.7 and 21.8).

(b) When secondary or tertiary datum features of size (with an axis or median plane) occur in the same tolerance frame and are interrelated by a different tolerance of

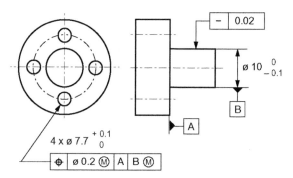

Fig. 21.6 Datum feature (B) with an axis and with a form tolerance that does not apply to the axis (MMVS for B Ⓜ is ø10)

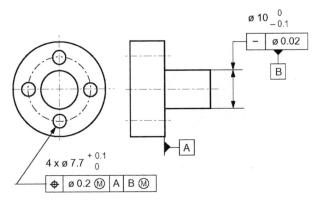

Fig. 21.7 Datum feature (B) with an axis and with a straightness tolerance to the axis (MMVS for B Ⓜ is ø10.02)

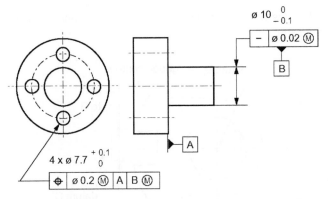

Fig. 21.8 Datum feature (B) with an axis and with a straightness tolerance to the axis and with the maximum material requirement to this straightness tolerance (MMVS for B Ⓜ is ø10.02)

orientation or location between each other, the maximum material virtual condition in the geometrical exact orientation and location relative to each other applies (Fig. 21.9). The maximum material virtual size of the datum is equal to the maximum material size plus or minus the corresponding tolerance of orientation or location (Fig. 21.10).

The datum virtual condition rule is not yet internationally standardized. However, part of it is self-evident (see 9.2). What is internationally disputed is that part of the rule dealing with geometrical tolerances to axes without the symbol Ⓜ (Fig. 21.7, see 9.2).

Figure 21.6 shows a case of rule (a). For the datum B the maximum material requirement Ⓜ applies. In addition, a form tolerance applies to the feature (generator line), not to the feature axis. For the gauge (to inspect the position of the four holes) at the datum B the virtual condition is equal to the maximum material condition (size ø10). This coincides with international practice.

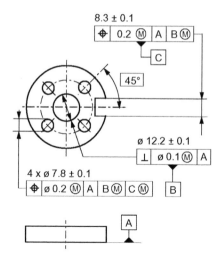

Fig. 21.9 Datums with Ⓜ, in the same tolerance frame and toleranced in relation to each other

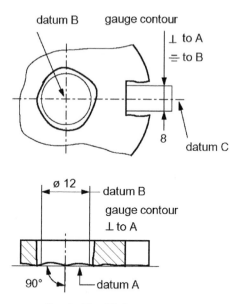

Fig. 21.10 Gauge according to Fig. 21.9

Figure 21.7 shows another case of rule (a). For the datum B the maximum material requirement Ⓜ applies. In addition, a straightness tolerance of the axis (without Ⓜ) applies. For the gauge (to inspect the position of the four holes) at the datum B the maximum material virtual condition of ø10.02 applies. This is internationally disputed. ISO will probably specify that the maximum material size (ø10) applies.

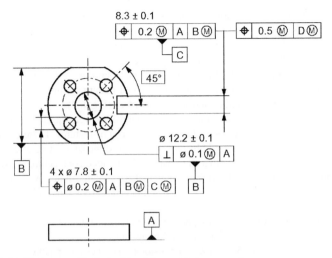

Fig. 21.11 Datums with Ⓜ, but in different tolerance frames

The same maximum material virtual condition of ø10.02 applies when the straightness tolerance of the axis is ø0.02 Ⓜ (Fig. 21.8). This coincides with international practice.

Figure 21.9 shows a case of rule (b). The datums A, B and C occur in the same tolerance frame (tolerancing the position of the four holes). The maximum material requirement applies to B and C. B is toleranced in relation to A (with Ⓜ). C is toleranced in relation to A and B (with Ⓜ). For the gauge (to inspect the position of the four holes) at the datums B and C the maximum material virtual conditions (ø12 and 8) apply (Fig. 21.10). This coincides with international practice.

Figure 21.11 shows a part similar to Fig. 21.9. However, there is an additional positional tolerance indicated at the datum feature C in relation to the datum D. Datum D does not occur in the tolerance frame of the positional tolerance of the four holes. Therefore the tolerance related to D is not to be taken into account for the calculation of the maximum material virtual condition of the datum C of the gauge (to inspect the position of the four holes). The gauge is the same as according to Figs 21.9 and 21.10. This coincides with international practice.

21.1.3 Positional tolerances according to ASME Y14.5

In the past, according to USASI Y14.5 – 1966 and according to ANSI Y14.5 – 1973, with positional tolerances, the maximum material requirement was applied to the toleranced features as well as to the datum features without any drawing indication of the maximum material requirement. This was applied if not otherwise indicated in the drawing (e.g. by Ⓢ or Ⓛ).

ANSI Y14.5M – 1982 required, with positional tolerancing, the indication Ⓜ, Ⓢ, or Ⓛ at the toleranced features (behind the positional tolerance) as well as at the datum features (behind the datum letter in the tolerance frame).

ASME Y14.5M – 1994 specifies with the new rule #2 that it also applies with positional tolerances regardless of feature size, if not otherwise specified with Ⓜ or Ⓛ.

According to ISO 1101, if not otherwise specified (i.e. without any specific indication), Ⓢ applies. This coincides with the new rule #2.

In the past, according to USASI Y14.5 – 1966 and according to ANSI Y14.5 – 1973, the following was applied to a positional toleranced pattern (group) of features (e.g. pattern of holes) with adjacent dimensional tolerances (Fig. 6.7).

The dimensional tolerances of the adjacent dimensions were applied to the centres of the positional tolerance zones of the pattern (group) of features. Therefore individual positional tolerance zones could be half outside the dimensional tolerance when the feature was at MMC, provided that the positional tolerance zone centres were within the dimensional tolerances. Therefore the actual axes of the holes were allowed to exceed the dimensional tolerances by half of the positional tolerance when the feature was at MMC (Fig. 6.7).

According to ISO 8015, this is not permitted. Dimensional tolerances as well as positional tolerances apply to the actual axes of the features.

According to ANSI Y14.5M – 1982 dimensional tolerances for positioning a pattern (group) of features that is positional toleranced between each other should be avoided and replaced by positional tolerances related to a datum system (Fig. 6.8). ASME Y14.5M – 1994 gives similar advice.

21.1.4 Composite tolerancing and single tolerancing

ASME Y14.5M – 1994 distinguishes between composite tolerancing and single tolerancing where there are two tolerances of the same type (position or profile) specified for the same features. Composite tolerancing is to be identified by one single tolerance

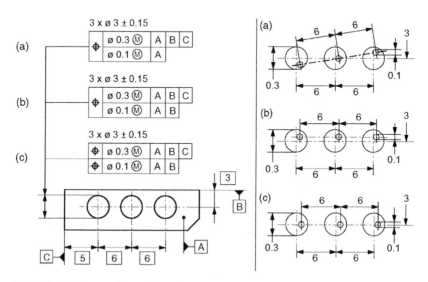

Fig. 21.12 Composite positional tolerancing (a) and (b); single positional tolerancing (c)

symbol for two tolerances, whereas single tolerancing is to be identified by separate tolerance symbols for each tolerance (Fig. 21.12).

In the case of composite tolerancing the upper (always larger) tolerance applies to the **location** of the pattern of features relative to the datums specified, and defines the Pattern-Locating Tolerance Zone Framework (PLTZF)*. The lower (always smaller) tolerance applies to the features relating to each other (interrelation of the features within the pattern) and, if applicable, to the **orientation** of the features as a group relative to the datums specified in the lower tolerance frame, and defines the Feature-Relating Tolerance Zone Framework FRTZF*. The theoretical exact dimensions from the datums (used to relate the PLTZF to the datums, TED 3 in Fig. 21.12) do not apply to the smaller tolerance (in the lower tolerance frame) that defines the FRTZF. The actual features (e.g. the actual axes) must lie within both tolerance zones (frameworks) (Fig. 21.12 (a) and (b)).

In the case of single tolerancing each tolerance applies as if no other tolerance were indicated, i.e. the theoretical exact dimensions from the datums specified in the lower tolerance frame apply also to the Feature-Relating Tolerance Zone Framework FRTZF. The actual features (e.g. the actual axes) must lie within both tolerance zones (frameworks) (Fig. 21.12 (c)).

ISO 1101 and ISO 5458 do not provide this differentiation. Also, when only one symbol is indicated for two tolerances, the indications of the theoretical exact dimensions relative to the datum(s) apply (as explained above for single tolerancing). See also Fig. 20.13.

21.1.5 Symmetry and concentricity tolerances

ANSI Y14.5M – 1982 did not specify the symbol $=$ for symmetry tolerances (it was only contained as former practice). Where it was required that a feature be located symmetrically with respect to the median plane of a datum feature, positional tolerancing (symbol $\oplus$) was to be used. According to ISO 1101 in these cases the symbol $=$ is used.

ASME Y14.5M – 1994 distinguishes between positional tolerances (with or without modifier M or L) and concentricity tolerances or symmetry tolerances (without modifier M or L).

Positional tolerances (symbol $\oplus$) are applied to the axis or median plane of the actual mating envelope (e.g. maximum inscribed cylinder for holes or minimum circumscribed cylinder for shafts orientated relative to the datum(s)) (Fig. 21.13).

Concentricity tolerances (symbol $\odot$) and symmetry tolerances (symbol $=$) are applied to the centres of the actual local sizes.

Figure 21.13 shows an actual workpiece that is within the positional tolerance but exceeds the concentricity tolerance.

ISO 1101 does not provide this distinction. According to ISO 1101 the tolerances of position, concentricity (coaxiality), symmetry, all without modifier, apply (all in the same way) to the actual axis or actual median face of the actual feature (see 3.5).

---

*The acronyms are pronounced "Fritz" and "Plahtz".

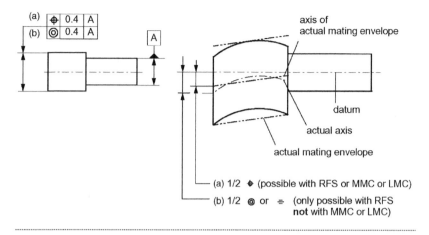

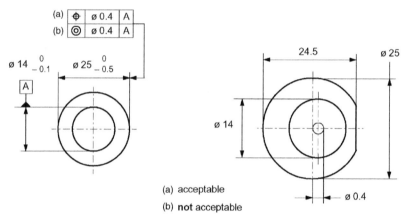

Fig. 21.13 Positional tolerance, concentricity tolerance, according to ASME Y14.5, top: extreme permissible workpiece below: workpiece within the positional tolerance but exceeding the concentricity tolerance

21.1.6 Orientation tolerances for axes or median planes

Orientation tolerances according to ASME Y14.5 apply to the (perfect) axes or median planes of the mating envelope (true geometric counterpart). Orientation tolerances according to ISO 1101 apply to the (imperfect) actual axes or median faces (Fig. 21.14).

21.1.7 Tangent plane requirement

ASME Y14.5M – 1994 introduced the tangent plane requirement to be specified by the symbol ⓣ after the orientation or location tolerance value. Then the tolerance zone applies to the high points of the controlled feature only (not to the entire surface irregularities) as shown in Fig. 21.5.

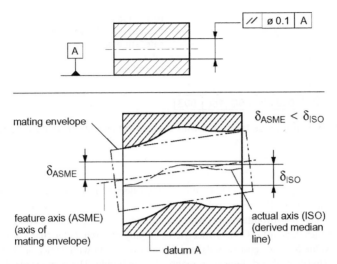

Fig. 21.14 Orientation (parallelism) tolerance according to ASME Y14.5 applied to the axis of the mating envelope and according to ISO 1101 applied to the actual axis

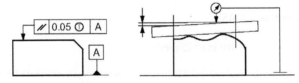

Fig. 21.15 Tangent plane requirement ⓣ

For inspection a contacting plate with sufficiently parallel planes may be used (Fig. 21.15). At present there is no rule for how to proceed in cases when the plate rocks about the highest points (e.g. when the controlled feature is convex).

21.1.8 Roundness measurements

ANSI B89.3.1 gives the possibility of specifying on drawings for roundness tolerances the measuring conditions, i.e.

- reference method, to be selected from:
  - MRS minimum radial separation
  - LSC least squares circle
  - MIC maximum inscribed circle
  - MCC minimum circumscribed circle
- filter, to be selected from:
  - 0, 1.67, 5, 15, 50, 150, 1500 cycles per revolution
- stylus tip radius, to be selected from:
  - 0.001, 0.003, 0.010, 0.030, 0.100 in

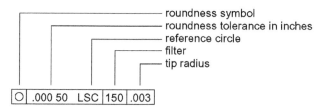

Fig. 21.16 Roundness tolerance with measuring conditions according to ANSI B89.3.1

See Fig. 21.16.

The filters are defined as electrical low-pass filters equivalent to two unloaded RC networks in series. Their nominal value (cut-off) corresponds to the sinusoidal frequency (cycles per revolution, cpr) whose amplitude is 70.7% transmitted by the filter.

The former standard ISO 4291 specified a slightly different filter characteristic. The nominal value (cut-off) corresponds to the sine wavelength with 75% amplitude transmission. However, this difference is practically negligible. The new standard ISO 12 181 will change the nominal value (cut-off) corresponding to the sine wavelength with 50% amplitude transmission by the Gauss filter.

ANSI B89.3.1 states further that, if not otherwise specified (default case), for roundness measurements (roundness tolerances) the following measuring conditions apply:

- reference method   MRS
- filter   50 cpr
- tip radius   0.010 in (0.25 mm)

The ISO Standards have not yet standardized measuring conditions for the default case. If necessary, they must be agreed upon between the parties (e.g. by reference to ANSI B89.3.1). See also 18.10.

21.1.9 Multiple patterns of features

According to ASME Y14.5M – 1994 positional toleranced multiple patterns of features (two or more patterns of features) located by theoretical exact dimensions relative to the same datum system (datums in the same order of precedence and modification) are regarded as a single pattern (simultaneous requirement), if not otherwise specified, e.g. with the notation "SEP REQT". This does not apply to the lower segment of composite tolerances (see 21.1.4) unless specific notation is added, e.g. "SIM REQT". ISO Standards do not specifically express this, see 20.6.5.

21.1.10 Terminology

Some terms given in ASME Y14.5 differ from those given by ISO (see Table 21.3).

ASME Y14.5M specifies: FIM Full Indicator Movement: the total movement of an indicator when appropriately applied to a surface to measure its variations. The former terms FIR Full Indicator Reading and TIR Total Indicator Reading have the same meaning as FIM.

Table 21.3 Comparison of ASME Y14.5 and ISO terminologies

ASME Y14.5M	ISO
Basic dimension	Theoretical exact dimension (TED)
Feature control frame	Tolerance frame
Variation	Deviation
True position (TP)	Theoretical exact position
Reference dimension	Auxiliary dimension

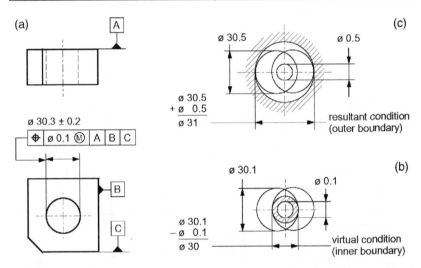

Fig. 21.17 Virtual and resultant condition boundaries, internal feature, MMC (a) drawing indication, (b) virtual condition, (c) resultant condition

ASME Y14.5M defines the terms "resultant condition", "inner boundary" and "outer boundary" (which are not yet used in ISO Standards) as follows.

Resultant condition: The variable boundary generated by the collective effects of a size feature specified MMC or LMC material condition, the geometric tolerance for that material condition, the size tolerance, and the additional geometric tolerance derived from the feature departure from its specified material condition, see Figs 21.17 to 21.20.

Inner boundary: A worst case boundary (that is, locus) generated by the smallest feature (MMC for an internal feature and LMC for an external feature) minus the stated geometric tolerance and any additional geometric tolerance (if applicable) from the feature's departure from its specified material condition, see Figs 21.17 to 21.20.

Outer boundary: A worst case boundary (that is, locus) generated by the largest feature (LMC for an internal feature and MMC for an external feature) plus the stated geometric tolerance and any additional geometric tolerance (if applicable) from the feature's departure from its specified material condition, see Figs 21.17 to 21.20.

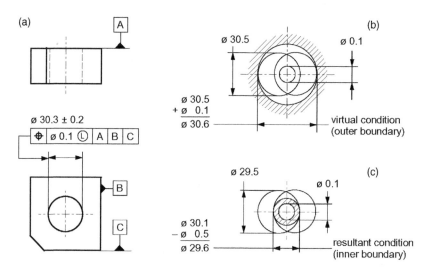

Fig. 21.18 Virtual and resultant condition boundaries, internal feature, LMC (a) drawing indication, (b) virtual condition, (c) resultant condition

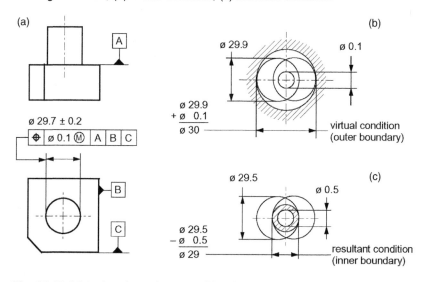

Fig. 21.19 Virtual and resultant condition boundaries, external feature, MMC (a) drawing indication, (b) virtual condition, (c) resultant condition

These terms may serve to assess the extent of a feature of size relative to its true (theoretical exact) position or to other features.

21.1.11 Survey of differences between ASME and ISO

Table 21.4 together with Tables 21.1 to 21.3 give a survey of the differences between ASME and ISO Standards.

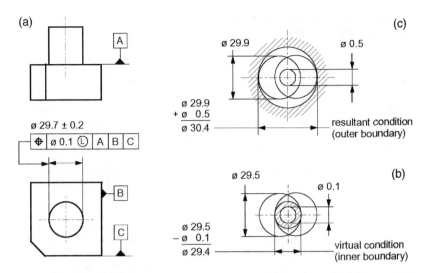

Fig. 21.20 Virtual and resultant condition boundaries, external feature, LMC (a) drawing indication, (b) virtual condition, (c) resultant condition

21.1.12 Drawings for both systems

21.1.12.1 General

Drawings for both systems, ISO and ASME, may be identified by the indication in the drawing title block:

ISO 1101, ISO 8015, ASME Y14.5 without rule #1

This gives the information where the meanings of the symbols are found. When the same symbol has different meanings in both systems, both meanings must be allowed by the function of the workpiece (and be accepted from the manufacturer) or the standard of the applicable meaning must be allocated to the drawing indication.

The following gives recommendations as to how to proceed in the relevant cases.

21.1.12.2 Envelope requirement

The use of the more economic principle of independency according to ISO 8015 is recommended. When the envelope requirement is necessary the symbol Ⓔ (ISO 8015) or straightness zero Ⓜ (ASME Y14.5) is to be indicated, see e.g. Fig. 20.50 (d) and (e).

21.1.12.3 Angular tolerances

Angular dimensional tolerances (Figs 21.3, 21.4 and 3.45) should be avoided or only be applied to less important cases, when the location of the angle and the form deviations need not to be toleranced.

Tolerancing of angularity, profile or position according to ISO 1101 is recommended (Figs 3.46, 20.14(b) and 20.19).

Table 21.4 together with Tables 21.1 to 21.3 give a survey of the differences between ASME and ISO Standards

Designation	ASME Y14.5	ISO
Envelope requirement Ⓔ Rule # 1	Ⓔ applies without drawing indication see 21.1.2	Ⓔ or 0 Ⓜ to be indicated or reference to ISO 2768- ... -E (or to DIN 7167) see 17.4 ff
Radius	surface to be contained within 3-D tolerance zone see 21.1.2	surface lines to be contained within 2-D tolerance zones (ISO TR 16 570) see 3.7
Angle	surface to be contained within 3-D tolerance zone see 21.1.2	contacting lines in section planes within tolerance (ISO TR 16 570) see 3.7
Datum at MMC	geom. tolerance of datum axis or median face without Ⓜ contributes to the gauge size see Fig. 21.7	geom. tolerance of datum axis or median face without Ⓜ does not contribute to the gauge size, see Fig. 9.3
Orientation tolerance of axis, median plane —————————— Positional tolerance	applies to the (straight) axis or (plane) median plane of the actual mating envelope see 21.1.5, 21.1.6	applies to the actual axis (not straight) or actual median face (not plane) see 21.1.5, 21.1.6
Symmetry tolerance concentricity tolerance (coaxiality tolerance)	applies to actual axis or actual median face (different from positional tol.) see 21.1.5	applies to actual axis or actual median face (same as positional tolerance) see 21.1.5
Composite tolerance	theor. exact dim. relative to the datum does not apply to the lower tolerance frame see 21.1.4	theoretical exact dimensions apply always unless otherwise specified (by wording)
Multiple pattern	default: SIM REQT see 21.1.9	to be indicated as applicable: simultaneous or separate
Roundness	default: MRS, 50 cpr, r = 0.01 in; see 21.1.8	default not yet standardized

21.1.12.4 Radii tolerances

Dimensional tolerancing of radii according to Fig. 21.1 should be avoided when the form deviations of the feature are important. The planned ISO Standard on these tolerances will probably differ from ASME Y14.5. In these cases profile tolerancing according to ISO 1101 is recommended.

21.1.12.5 Datum virtual condition rule

It is recommended to avoid the problematic case of Fig. 21.7 (datum feature axis or median surface geometrically toleranced without Ⓜ or Ⓛ).

21.1.12.6 Simultaneous requirements

It is recommended always to indicate "simultaneous" when it applies, see 21.1.9.

21.1.12.7 Orientation tolerances of axes or median faces

ISO 1101 tolerances the actual (imperfect) axes or median faces. ASME Y14.5 tolerances the (perfect) axes or median planes of the mating envelopes, see 21.1.6.

It is recommended to tolerance in a way that both versions fit to the workpiece function and are acceptable.

21.1.12.8 Positional tolerancing

It is recommended to avoid positional tolerancing of planar surfaces and to use profile tolerancing, as it is the rule according to ASME Y14.5.

ISO 1101 tolerances with positional tolerances the actual (imperfect) axes or median surfaces. ASME Y14.5 tolerances with positional tolerances the (perfect) axes or median planes of the mating envelopes, see 21.1.5 and Figs 21.13 and 21.14.

It is recommended to tolerance in a way that both versions fit to the workpiece function and are acceptable.

21.1.12.9 Composite tolerancing

It is recommended to indicate always 'orientation only' when it applies, see 21.1.4 and Fig. 20.13.

21.1.12.10 MMR, LMR

For tolerances with MMR or LMR it is recommended to use position and to avoid coaxiality and symmetry, as it is the rule according to ASME Y14.5.

21.2 BS 308 Part 2 – 1985

21.2.1 Tolerancing principles

The former British Standard BS 308 Part 2–1985 provided both concepts, principle of independency and principle of dependency of size and form with equal status. When the principle of independency was applied the drawing was to be marked with the symbol

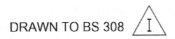

DRAWN TO BS 308

Otherwise the principle of dependency was applied (see 17.6).

21.2.2 Composite positional tolerancing

When two different positional tolerances were applied to the same features (Fig. 6.8), according to BS 308 Part 2 – 1985, the larger positional tolerance was regarded as the tolerance of the location of the features as a group (location of the pattern), while the smaller positional tolerance was regarded as the tolerance of the individual locations

Fig. 21.21 Positional tolerances (a) applied to the centres of (other) positional tolerance zones (b) applied to the actual axes only

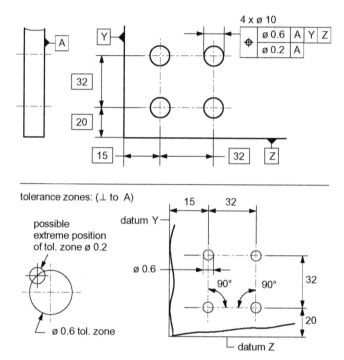

Fig. 21.22 Composite positional tolerances according to BS 308, the actual axis may utilize the full zones ø0.2

(location of the features relative to each other). Hence it was specified that the larger positional tolerance be applied to the centres of the tolerance zones of the smaller positional tolerance. Therefore the actual axis was allowed to exceed the larger tolerance zone by half of the smaller positional tolerance zone (Fig. 21.22).

According to ISO 1101 and ISO 5458 both positional tolerances apply to the actual axes. There is no exceeding of the larger positional tolerance allowed for the actual axis.

When applying ISO 1101, in order to express the same meaning as according to BS 308, the larger positional tolerance may be modified by the indication "axes of tolerance zones" (Fig. 21.21 (a)), or the larger positional tolerance is to be increased by the smaller tolerance value (Fig. 21.21 (b)).

21.2.3 Positional tolerances adjacent to dimensional tolerances

In contrast to ISO 1101 and ISO 5458, according to BS 308 Part 2 – 1985, the same was applied as to ANSI Y14.5 – 1973 (as described in 6.6.1), i.e. when there were positional tolerances adjacent to dimensional tolerances, the actual axis was allowed to exceed the dimensional tolerance by half the positional tolerance (Fig. 6.7).

21.3 DIN 7167 and former practices according to DIN 406 and DIN 7182

The International Standards ISO 1101, ISO 5458, ISO 5459, ISO 3040, ISO 2692 and ISO 8015 have been adopted as DIN Standards (German National Standards) without any alteration.

In addition, DIN 7167 has been issued. It specifies the principle of dependency for drawings produced according to DIN Standards without an indication of the principle of independency (without indication of ISO 8015 or DIN 2300). This applies to size and form of isolated (individual) features only (not to related features), as described in 17.6. It does not apply to features to which a form tolerance larger than the size tolerance applies (e.g. to stock material).

The former German Standard DIN 406 had standardized a drawing indication as shown in Fig. 21.23. It specified the maximum permitted separation of axes or median faces. The equivalent indication according to ISO 1101 is shown in Fig. 21.23.

The former German Standard DIN 7182 Part 4 March 1959 symbolized the permissible roundness deviation or the permissible cylindricity deviation by word indication as shown in Fig. 21.24. In this case the tolerance was diameter related (not radius related as in ISO 1101). The equivalent indication according to ISO 1101 is shown in Fig. 21.24.

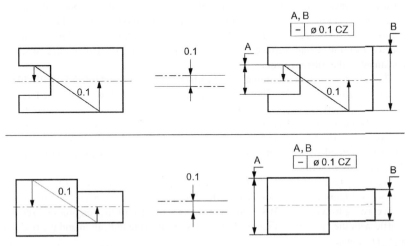

Fig. 21.23 Permissible eccentricity according to DIN 406 – 1968 and the equivalent indication according to ISO 1101

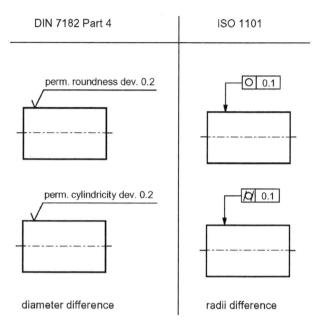

Fig. 21.24 Permissible roundness and cylindricity deviations according to DIN 7182 Part 4 – 1959 and the equivalent indication according to ISO 1101

21.4 Former East European Standards

According to the former Comecon Standards ST RGW 368 – 76 it was to be distinguished between radius-related and diameter-related tolerance indications (Table 21.5).

According to ISO 1101 tolerances of orientation or location also limit the form deviations of the toleranced features (surfaces or axes or median faces). These standards specify that the actual surface or the actual axis or actual median face shall be contained in the tolerance zone.

This was not applied according to the former Comecon Standards ST RGW 301 – 76 and ST RGW 368 – 76. These standards specified that the orientation or location tolerance zone applies to the geometrical ideal contacting element (Fig. 18.14) or to the axis of the geometrical ideal contacting cylinder. The contacting element or contacting cylinder was orientated according to the minimum rock requirement. With holes, the contacting cylinder was the maximum inscribed cylinder while with shafts it was the minimum circumscribed cylinder.

According to ST RGW 368 – 76, there was a crossing tolerance of two axes with the symbol X. Figure 21.25 shows the drawing indication and the relevant tolerance zone.

According to ST RGW 368 – 76 there was a tolerance of the longitudinal section profile with the symbol =. Figure 21.26 shows the drawing indication and the relevant tolerance zone.

This tolerance considers generator lines of cylindrical features in sections containing the axis of the contacting cylinder. In these sections the tolerance zones were composed of two parallel pairs of parallel straight lines each the tolerance value apart. For

Table 21.5 Tolerance indications according to the former East European Standards

Tolerance	Former practice RS 430–65 TGL 19 085	ST RGW 368–76 TGL 31 049 radius related	diameter related	ISO 1101 diameter related
Coaxiality	⌐ 0.1	◎ R 0.1	◎ ø 0.2	◎ ø 0.2
Symmetry	÷ 0.1	≡ T/2 0.1	≡ T 0.2	≡ ø 0.2
Position of axis	+ 0.1	⊕ R 0.1	⊕ ø 0.2	⊕ ø 0.2
Position of plane or straight line	—	⊕ T/2 0.1	⊕ T 0.2	⊕ ø 0.2
Crossing of two axes	✕ 0.1	✕ T/2 0.1	✕ T 0.2	—

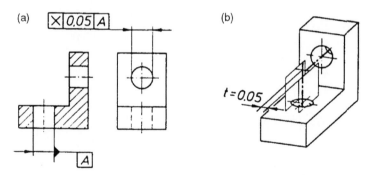

Fig. 21.25 Crossing tolerance of two axes: (a) drawing indication; (b) tolerance zone

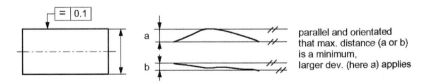

Fig. 21.26 Tolerance of the longitudinal section profile: (a) drawing indication; (b) tolerance zone

measurement the pair of parallel straight lines was to be orientated in such a way that the greater of the two distances within the pairs of parallel straight lines becomes a minimum.

With the deviation of the longitudinal section profile, the cylindricity deviation can be assessed by a close approximation. According to Heldt (see Ref. [12]) the cylindricity deviation δ_z is approximately equal to the roundness deviation δ_r plus the longitudinal section profile deviation δ_q

$$\delta_z \geqslant \delta_r + \delta_q$$

22

Tolerancing of Edges

For tolerancing the geometry (sharpness) of edges (within cross-sections) a separate standard ISO 13 715 has been developed. Edges are intersections of two surfaces. Edges always exhibit deviations from the ideal geometrical shape (burr or undercut or passing) (Fig. 22.1). The size of these deviations may be necessary to be toleranced for functional reasons (e.g. in hydraulic equipment) or out of safety considerations.

Burr is a rough remainder of material outside the ideal geometrical shape of an external edge, or a residue of machining or of a forming process (Fig. 22.1).

Undercut is a deviation inside the ideal geometrical shape of an internal or external edge (Fig. 22.1).

Passing is a deviation outside the ideal geometrical shape of an internal edge (Fig. 22.1).

The size a of the edge deviation is illustrated in Fig. 22.2.

The tolerances for the deviations from the ideal geometrical shape of edges within cross-sections are to be indicated by the symbol shown in Fig. 22.3.

The indication + at the symbol means only deviations outside the material of geometrical ideal form are allowed (burr or passing permitted, undercut not permitted) (Fig. 22.5).

The indication − at the symbol means only deviations inside the material of geometrical ideal form are allowed (undercut permitted, burr or passing not permitted) (Fig. 22.5).

The indication ± at the symbol means deviations outside and inside the material of geometrical ideal form are allowed (undercut or burr or passing permitted).

When necessary the upper limit of the size of the deviation or the upper and lower limit of the size of the deviation is to be indicated following the sign (Fig. 22.5). A sharp edge is an edge with an indication of a very small edge tolerance (e.g. −0.05 mm).

When it is needed to specify the direction of the burr on an external edge or the undercut on an internal edge this shall be indicated as shown in Fig. 22.3 at place a_2 or a_3 and in Fig. 22.4.

When the symbol points to a corner (point) the edge vertical to the projection plane (front view) is meant or the edge of the feature (e.g. hole) is meant (Fig. 22.5).

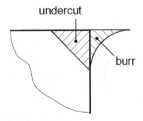
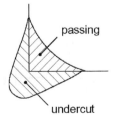

Fig. 22.1 Edge deviations

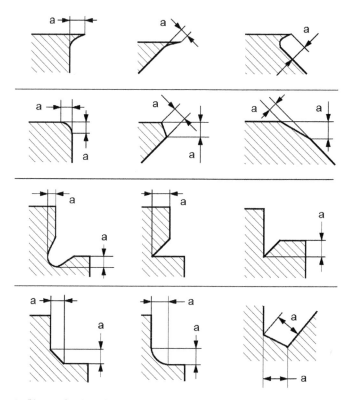

Fig. 22.2 Sizes of edge deviations

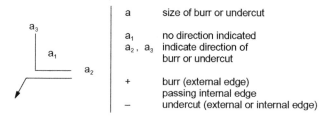

a	size of burr or undercut
a_1	no direction indicated
a_2, a_3	indicate direction of burr or undercut
+	burr (external edge) passing internal edge
−	undercut (external or internal edge)

Fig. 22.3 Symbol for edge tolerances

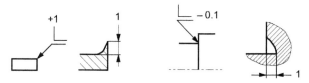

Fig. 22.4 Indication of the direction of the burr on an external edge and of the undercut on an internal edge

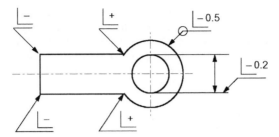

Fig. 22.5 Examples of indication of an edge tolerance applied to all edges around the profile on both sides, front and back; applied to the edge of a feature (hole); applied to edges vertical to the projection plane and without tolerance value

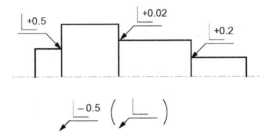

Fig. 22.6 Example of indication of a collective edge tolerance (−0.5)

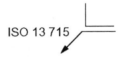

Fig. 22.7 Reference to ISO 13 715

When the symbol points onto a line this edge is meant; if only one view is represented and the outlines of both front and back are the same the edges of the front and the back are meant (Fig. 22.5).

When the symbol "all around" is used all edges around the profile are meant (Fig. 22.5). In the case of ambiguity, the indication may be used at corners. The symbol "all around" shall not be used in sectional representations.

The edge tolerance indication may be applied

- as an individual indication for a single edge (Figs 22.5 and 22.6);
- as an individual indication for all edges around the represented profile (Fig. 22.5);
- as a collective indication near the title block common to all edges without individual edge tolerance indication (Fig. 22.6).

According to ISO 13 715 without an indication of the state of the edges, the parts may be delivered direct from the machine without an additional edge treatment.

When the symbols and rules according to ISO 13 715 (described above) are applied, it is recommended that reference be made to ISO 13 715 either within or near the title block or within allocated documents, in the manner shown by Fig. 22.7.

23

ISO Geometrical Product Specifications (GPS), New Approach

23.1 Terms and definitions

ISO aims to give mathematically exact definitions of the constituents of the geometrical specifications in order to give a sound foundation for

- unambiguous drawing specifications;
- programming of measurement instruments;
- estimation of measurement uncertainties.

All geometrical specifications deal with **features**. Their definitions were an issue of intensive discussions within ISO.

According to ISO 14 660-1, ISO TS 17 450-1 and ISO 22 432-1:

(Single) feature: geometric entity which is a single point, a single line, or a single surface, e.g. a cylinder.

Compound feature: geometric entity which is a collection of features, e.g. two parallel opposite planes connected by a size dimension.

According to ASME Y14.5:

Feature: physical portion of a part, such as a surface, pin, tab or slot.

The meaning of a drawing indication remains the same despite the different definitions of a feature.

ISO 14 660-1 defines

- **nominal feature** (of ideal geometry according to the drawing);
- **real feature** (of non-ideal geometry as existing on the workpiece);
- **extracted feature** (of non-ideal geometry as detected from the workpiece);
- **associated feature** (of ideal geometry and fitted to the extracted feature according to an objective function (fitting rule, e.g. minimum zone));

see Fig. 23.1.

ISO 22 432 Parts 1 to 6 further describe various types of feature:

- **ideal feature** and **non-ideal feature;**
- **integral feature** (ideal or non-ideal, surface or line on a feature) and **derived feature** (centre point, median line or median surface derived from one or more integral features);

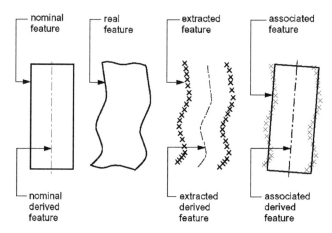

Fig. 23.1 Features

- **feature of size** (e.g. hole or shaft (single feature), tab or slot (compound feature));
- **nominal feature** (ideal, defined by the drawing) and **specification feature** (non-ideal, defined by the specification operator) and **verification feature** (non-ideal, defined by the verification operator);
- **filtered feature** (e.g. feature after Gauss filtering);
- **associated feature** (ideal feature or set of ideal features fitted to the feature, e.g. smallest circumscribed cylinder, V-block);
- **candidate features** (set of ideal features which satisfies geometrical constraints like outside the material used to model the function of a fit);
- **substitute feature** (unique ideal feature associated with a non-ideal feature, e.g. as used in CMM techniques);
- **limited feature** (portion of an ideal feature, e.g. restricted area);
- **enabling feature** (ideal feature, identified from an ideal feature and used solely to build other features, e.g. section plane).

ISO 17 450-1 defines the **duality principle** by defining the **non-ideal surface model** of the workpiece (skin model, imagination of the designer, design intent) and the **verification model** (defined by the verification process executed by the inspector on the manufactured workpiece). The difference between both models leads to the measurement uncertainty.

The models incorporate the following **operations**:

- **partitioning** (application part, area);
- **extraction** (assessment of points of the workpiece);
- **filtration** (e.g. probing by a sphere, using a Gauss filter);
- **association** (fitting an ideal feature to the non-ideal extracted and filtered feature by using a particular method (objective function), e.g. Gauss or Chebyshev or maximum inscribed feature or minimum circumscribed feature);
- **collection** (taking together, e.g. two coaxial cylinders as a common datum);

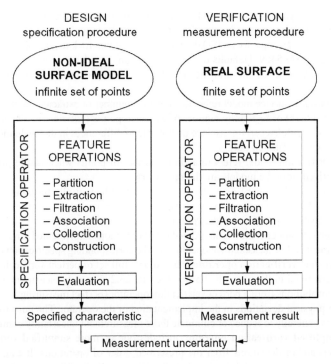

Fig. 23.2 Duality principle

- **construction** (e.g. building a straight line as a datum (situation feature) by the intersection of two associated planes in the case of a wedge);
- **evaluation** (determining the value of the characteristic (e.g. diameter) or the deviation);

see Fig. 23.2.

The operation is a tool required to obtain a feature or the value of a characteristic. Each operation that may influence the result (feature or value of a characteristic) should be specified by an indicated specification or by the use of a default.

For example, the measurement result of a diameter of a shaft measured by a coordinate measuring machine depends on the following operations:

- number of points probed (extraction);
- stylus tip radius (extraction, filtration);
- filter, if any (filtration);
- adjustment of the workpiece to the reference, e.g. minimum circumscribed cylinder (association);
- algorithm used for calculation, e.g. circumference divided by π (evaluation).

Defaults are standardized specifications of operators to be used when no specification is indicated (applies if not otherwise specified).

When GPS symbols are used and no reference to other standards (e.g. to a national standard) is indicated on the drawing the ISO GPS standards apply. Then the ISO defaults apply, if any. An example for an ISO default is the minimum rock requirement for datums according to ISO 5459. Defaults defined by company standards or national standards must be referred to on the drawing; in which case they overrule the pertinent ISO defaults.

The non-ideal surface model refers to an infinite number of extracted points (filtering only by a sphere) or to a finite number of extracted points (filtering by a Gauss filter, where points within a certain distance do not lead to differences in the evaluation result because they are filtered out).

The verification model refers to a finite number of points when tactile probing is used. Derived from the models are the **operators**.

The **specification operator** is an ordered set of (mathematically formulated) operations, i.e. a theoretical concept to identify a requirement to a feature (feature attributes), using the feature operations as shown in Fig. 23.2.

The **verification operator** is an ordered set of operations implemented physically in a measurement and/or measurement operation of the corresponding specification operator.

The **perfect verification operator** is based on a full set of perfect verification operations performed in the prescribed order, see Fig. 23.2. The only measurement uncertainty contributions from a perfect verification operator (perfect measuring instrument) are from physical deviations in the implementation of the operator. The purpose of calibration is generally to reduce or eliminate these measurement uncertainty contributors.

A **simplified verification operator** includes one or more simplified verification operations and/or deviations from the prescribed order of operations. It causes additional measurement uncertainty contributions. The magnitude of these uncertainty contributions is dependent on the geometrical characteristics (geometrical deviations) of the actual workpiece. For example, the procedure for measurement of roundness deviations using a V-block and dial gauge is a simplified verification operator. When the workpiece has no lobed form deviation this does not contribute to the measurement uncertainty.

ISO TS 17 450-2 defines the following **types of uncertainty**:

Specification uncertainty due to a specification which is not correctly defined (inherent uncertainty by not complete and/or not mathematically correctly defined operations); e.g. step dimension with ± tolerance.

Correlation uncertainty arising from the difference between the specification (operator) and the (real) functional requirement.

Method uncertainty arising from the difference between the (actual) specification operator and the (actual) verification operator, disregarding the physical deviations of the actual verification operator.

Implementation uncertainty arising from the physical deviations of the verification operator.

Measurement uncertainty arising from the difference between the specification operator and the verification operator (including those contributions caused by physical deviations in the implementation of the operator). It is equal to the sum of the method uncertainty and the implementation uncertainty. According to GUM it is a parameter, associated with the result of a measurement, that characterizes the dispersion of the values that could reasonably be attributed to the measurand (particular quantity subject to measurement).

ISO TS 17 450-2 refers to GUM and VIM. Accordingly the measurement uncertainty is expressed in a statistical way. According to ISO 14 253-1 as the expanded uncertainty U (of a measurement) with a coverage factor 2 to the standard deviation u, if not otherwise indicated, i.e. $\pm U = \pm 2u$.

How to deal with measurement uncertainty is given by the conformance rules (decision rules), see 18.11.2.

23.2 Filters

Filters are used to assess particular components of the surface, e.g. when roundness (circularity) deviations are to be assessed the roughness shall be eliminated by filtering. Otherwise the measuring result may be significantly falsified by the roughness, e.g. a turned perfect circular cylinder would show a roundness deviation of the roughness depth, see Fig. 23.3.

There are various filter types available, see ISO TS 16 610-1 ff and the following table. The filter symbols are intended for drawing indications.

There is no default filter standardized for GD&T. When not otherwise indicated mainly morphological filters (sphere filtering) are used and no other filtration. When using dial gauges the usual sphere diameter is 3 mm. For very rough surfaces larger sphere diameters may be necessary. For smaller holes and slots smaller sphere diameters or auxiliary devices (like gauge blocks or mandrels) may be used.

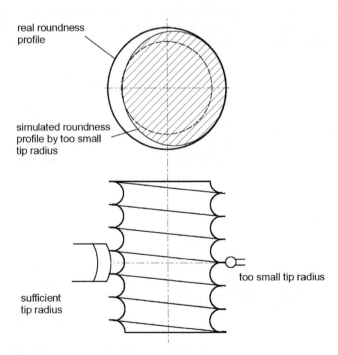

real roundness profile

simulated roundness profile by too small tip radius

too small tip radius

sufficient tip radius

Fig. 23.3 Roughness falsifies form deviation

The **nesting index** defines the filter size. It is similar to the grid size of a mesh sieve. For morphological filters the nesting index is given by the sphere radius, for Gauss and spline filters it is given by the cut-off (e.g. in the case of surface roughness). A discrete surface model consists of a finite number of points. When the nesting index is low the number of required points of the model is high. When the nesting index tends to zero the discrete surface model tends towards the skin model of an infinite number of points.

Morphological filters (other names are sphere filters and ball filters) retain the high points of the surface and eliminate narrow valleys into which the sphere cannot penetrate, see Fig. 23.4. These filters may be used for datums and fits.

Gauss filters and spline filters separate high and low frequency components (short and long wavelength components) of the surface. Figure 23.5 shows the components of the surface separated by different filter sizes (nesting indexes). (In the case of a ball bearing the deviations of median wavelength are more detrimental than the others and should be toleranced separately and very tightly.)

Gauss and spline filters may eliminate the high points of the surface, see Fig. 23.6. Therefore they are not suitable for datums and fits.

Spline filters work in a similar way to Gauss filters but do not have some of the Gauss filter's disadvantages (end effects, sometimes the mean line is out of the material, see ISO TS 16 610-1). Spline filters, however, are not yet popular.

Robust filters do not create distortions at outliers. RC filters and non-robust Gauss and spline filters do and falsify the measurement result.

Spline wavelet filters allow to detect outliers.

An overview with advantages and disadvantages of the various filters is given in ISO TS 16 610-1. The following shows the filters, their planned symbols for drawing indications, and the planned ISO document nos.

Arial filter

FALG	Gaussian	ISO TS 16 610-61
FALS	Spline	ISO TS 16 610-62
FALW	Spline wavelet	ISO TS 16 610-69
FAMCB	Closing ball	ISO TS 16 610-81
FAMCH	Closing horizontal segment	ISO TS 16 610-81

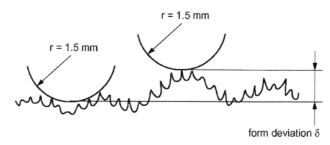

Fig. 23.4 Morphological filter, roughness filtered out, waviness and form deviations retained

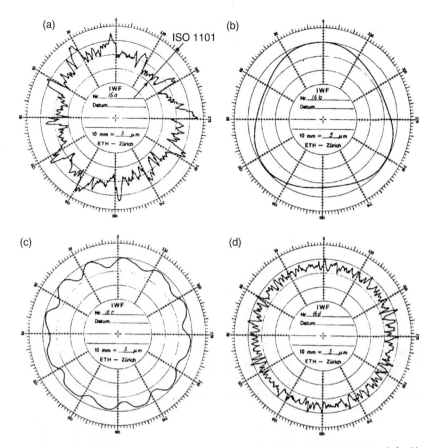

Fig. 23.5 Roundness deviation (same trace), (a) without filter (ISO 1101 default), (b) to (d) with different filters (source: Wirtz, ETH Zuric)

FAMOB	Opening ball	ISO TS 16 610-81
FAMOH	Opening horizontal segment	ISO TS 16 610-81
FAMAB	Alternating series ball	ISO TS 16 610-89
FAMAH	Alternating series horizontal segment	ISO TS 16 610-89
FARG	Robust Gaussian	ISO TS 16 610-71
FARS	Robust spline	ISO TS 16 610-72

Profile filter

FPLG	Gaussian	ISO TS 16 610-21
FPLS	Spline	ISO TS 16 610-22
FPLW	Spline wavelet	ISO TS 16 610-29
FPMCB	Closing ball	ISO TS 16 610-41
FPMCH	Closing horizontal segment	ISO TS 16 610-41

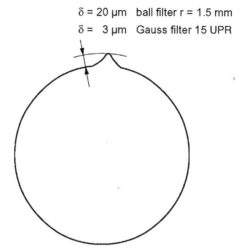

δ = 20 μm   ball filter r = 1.5 mm

δ =   3 μm   Gauss filter 15 UPR

Fig. 23.6 Gauss and spline filter do not assess the full height of narrow form deviations

FPMOB	Opening ball	ISO TS 16 610-41
FPMOH	Opening horizontal segment	ISO TS 16 610-41
FPMAB	Alternating series ball	ISO TS 16 610-49
FPMAH	Alternating series horizontal segment	ISO TS 16 610-49
FPRG	Robust Gaussian	ISO TS 16 610-31
FPRS	Robust spline	ISO TS 16 610-32
F2RC	2RC	ISO 3274

Symbols: F Filter, A Arial, P Profile, L Linear, M Morphological

There are still ongoing investigations to find out the appropriate applications of the filters according to the functional needs of the workpieces.

23.3 Datums

For the definition of datums ISO TS 17 450-1 and the planned new ISO 5459 refer to the **invariance classes** of ideal features (geometrical elements). All ideal features belong to one of the following seven invariance classes.

Each feature has 6 degrees of freedom (translations along the axes x, y, z and rotations around the axes x, y, z). The invariance class corresponds to the degrees of freedom (which are remaining and not locked). It describes the displacement of the feature (translation, rotation) for which the feature is kept identical in space.

Invariance class	Invariance degrees (remaining degrees of freedom)
complex	none
prismatic	1 translation
revolute	1 rotation
helical	combined translation and rotation
cylindrical	1 translation and 1 rotation
planar	1 rotation and 2 translations
spherical	3 rotations

The **situation feature** is a point, straight line, plane or helix, which defines the location and/or orientation of an ideal feature (datum). Examples are given below.

Invariance class	Example	Situation features	Invariance degrees (degrees of freedom)
complex	hyperbolic paraboloid	symmetry planes, tangent point	none
prismatic	prism with elliptic basis	symmetry planes, axis	1 translation along axis
revolute	cone	axis, apex	1 rotation around axis
	torus	plane, centre point	1 rotation around axis
helical	helical surface	helix	combined translation and rotation
cylindrical	cylinder	symmetry axis	1 translation along axis, 1 rotation around axis
planar	plane	plane	2 translations within plane, 1 rotation
spherical	sphere	centre point	3 rotations around centre

Screw threads are considered to belong to the invariance class cylindrical (because the pitch, i.e. the movement along the axis, is of no interest for the datum).

The **datum specification** refers to a situation feature. The situation feature derives from an associated ideal feature to the extracted feature. If not otherwise specified, the planned objective function for the association is minimum zone (=min.-max. requirement, Chebyshev) and the constraint is tangent to the material. Filtration other than with a probing sphere is not involved (planned new ISO 5459).

Theoretically other association methods (objective functions) can be specified, e.g. Gauss.

When the drawing indication for the datum in the tolerance frame contains the datum letter only, all situation features are applied. When only one situation feature is to be applied, e.g. the apex of the cone and not the axis, then the applied situation feature in parentheses is to be indicated, e.g. [PT] for the apex of the cone, see Fig. 23.7.

Planned drawing indications for selected situation features:

point	drawing indication: [PT]
straight line	drawing indication: [SL]
plane	drawing indication: [PL]

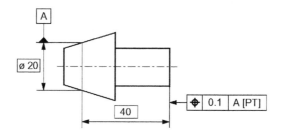

datum is the point (PT) of the cone axis
perpendicularity relative to cone axis not covered

Fig. 23.7 Selected situation feature point

23.4 Principle of independency and Ⓜ or Ⓛ

With the new approach a rigorous application of the principle of independency is planned.

For example, in Fig. 20.29 the indication "separate" is therefore not necessary. It applies according to the principle of independency. That is the opposite of the rule in ASME Y14.5, see 20.6.5 and 21.1.9. However, the author of this book recommends to indicate "separate", at least for a transition period and in order to avoid misunderstandings with ASME.

In Fig. 23.8, according to the principle of independency, each hole must fit to a gauge with two shafts separately (the holes need not fit to a gauge with seven shafts). According to former practice (and still at the time of publication of this book) it

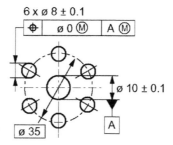

Fig. 23.8 Pattern of holes related to a datum hole with Ⓜ, CZ not indicated

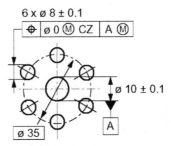

Fig. 23.9 Pattern of holes related to a datum hole with Ⓜ, CZ indicated

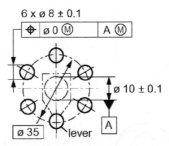

Fig. 23.10 Pattern of holes related to a datum hole with Ⓜ, CZ, not indicated, lever turning around the datum hole and fitting into each circumference hole separately

was usually assumed that simultaneous gauging (gauge with seven shafts) is to be applied. According to the principle of independency for simultaneous gauging "CZ" must be indicated after Ⓜ in the tolerance compartment in the tolerance frame (Fig. 23.9).

In the case of Fig. 23.8 separate gauging may be appropriate to a function where a lever with two shafts turns around the centre hole and must fit with the other shaft into each of the other holes (Fig. 23.10).

Figures 23.11 and 23.12 show examples where both possibilities ((a) simultaneous gauging and (b) separate gauging) are explained.

The indication of "CZ" is only relevant when a datum with Ⓜ is involved. Without Ⓜ there is no floating of the datum and it comes to the same result irrespective whether separate gauging (gauge with two shafts), each at the correct location or simultaneous gauging (gauge with seven shafts) is applied.

A similar situation appears when multiple features are toleranced with Ⓜ but without datum. Figure 23.13 shows an example. "CZ" means simultaneous gauging so that the box end wrench fits. Without "CZ" separate gauging is allowed so that the open end wrench fits but the box wrench does not.

In the case of separate gauging the relationship between the toleranced features must be toleranced in addition, e.g. by positional tolerancing without Ⓜ, (Figs 23.14 and 23.15).

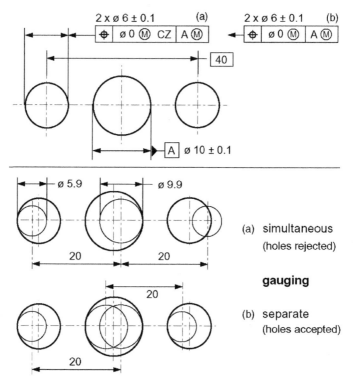

Fig. 23.11 Pattern of holes related to a datum hole with Ⓜ : (a) simultaneous gauging, (b) separate gauging

In order to avoid misunderstandings the author recommends the following (at least for a transition period):

For designer:	To indicate "CZ" when simultaneous gauging is necessary, e.g. Fig. 23.9 when the function is fastener assembling
	To indicate "single" (as in Fig. 23.16) when separate gauging is allowed.
For manufacturer:	Manufacture always according to "CZ" (also when "CZ" is not indicated) when "single" is not indicated or ask whether separate gauging is allowed.

Similar situations occur with the least material requirement Ⓛ .

23.5   Conformance rules

ISO 14 253-1 gives conformance rules, i.e. how to decide whether a workpiece conforms to the specification or not.

According to this standard the default is that the measurement uncertainty goes to the debit of the measuring party, see Fig. 23.17.*

---

* Intermediates (merchants) should not inspect, they should use the manufacturer's inspection documentation.

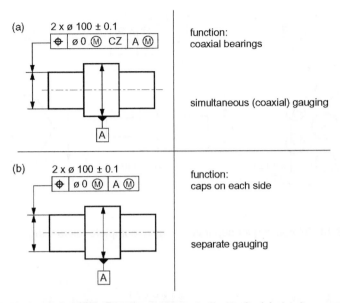

Fig. 23.12 Coaxial shafts related to a datum shaft with Ⓜ : (a) simultaneous gauging, (b) separate gauging

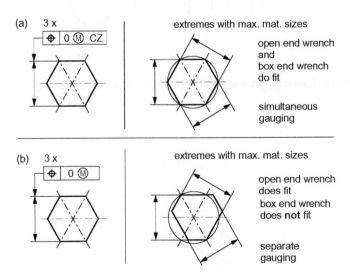

Fig 23.13 Hexagon, widths across flats with Ⓜ: (a) simultaneous gauging, (b) separate gauging

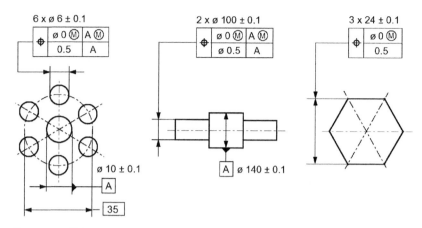

Fig. 23.14 Tolerancing for separate gauging, complete tolerancing

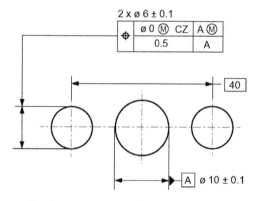

Fig. 23.15 Tolerancing for separate gauging, complete tolerancing

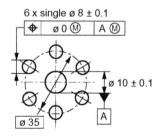

Fig. 23.16 Pattern of holes related to a datum hole with Ⓜ, indication of "single" for separate gauging

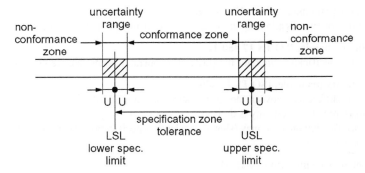

Result of measurement within uncertainly range:
three possible solutions:

(1) Lower uncertainty of measurement until result of measurement outside uncertainly range
(2) In advance agreement how to deal with
(3) Rejection for outgoing inspection. Acceptance for incoming inspection

Fig. 23.17 Measurement uncertainty and decision rule

This means that a workpiece cannot be rejected when the measurement exceeds a tolerance by an amount smaller than the measurement uncertainty, see Fig. 23.17.

This means that possibly in reality the tolerance has been exceeded, but the workpiece cannot be rejected. This may be acceptable in many cases, but in some cases it is not. Therefore another rule must be agreed upon, e.g. the following:

In cases of the maximum material requirement Ⓜ the drawing specifies the absolute functional limit. For the contract a contractional tolerance may be applied. The contractional tolerance is at the maximum material limit of 10% within the drawing (functional) tolerance. The 10% complies with the measurement uncertainty of the purchaser. Obviously, instead of 10% any other value according to the purchaser's measurement uncertainty can be agreed upon.

See also 18.11.

23.6 New approach principles and rules

The following principles and rules are adopted with the new approach:

Independency principle, ISO 8015
(applied at present when "ISO 8015" is indicated in drawing title box, intended by ISO TC 213 to become the default rule without indication)
Each requirement must be indicated and respected (no hidden rule as for example the envelope requirement must be respected for each feature of size).

Drawing definitive principle, ISO 14 659
What is not specified is not required. The specification may be directly indicated or be given by the reference to a default (ISO standard default, national standard default, company standard default).

Operator principle, ISO TS 17 450-1&2

Definition of specifications by an ordered set of operations, see Fig. 23.2.

Duality principle, ISO TS 17 450-1&2

From the specification operator the verification operator is derived, see Fig. 23.2.

Uncertainty principle, ISO TS 17 450-2

From the difference between the specification operator and the verification operator, the measurement uncertainty is derived.

The specification operator may (should not) have a specification uncertainty and a correlation uncertainty.

Conformance rules, ISO 14 253-1

Measurement uncertainty goes to the debit of the measuring party, if not otherwise specified.

Decimal rule, ISO 14 659

After the last indicated digits of dimensions and tolerances, zero applies (no rounding).

Reference temperature, ISO 1

20°C applies.

Specification interpretation rule, ISO 14 659

Applied is the standard in force at the time of drawing (specification) issue.

23.7 Application of the new approach

The new approach with its sophisticated terms and definitions is relevant for

- programming of measuring instruments (like coordinate measuring machines);
- definition of very small tolerances;
- precise evaluation of the measurement uncertainty.

For normal designer use (geometrical tolerances larger than 20 μm) the terminology as used in 1 to 22 is considered as sufficient (and easier to understand).

When very small tolerances are required, the precise definition of the operators (filter etc.) according to the new approach may be necessary. For example, for the piston of the injection pump of a diesel engine, where tolerances of 1 μm or less are required for roughness, waviness, roundness and size.

In the existing ISO standards, there are no standardized default (=if not otherwise stated) conditions

- for the extraction (number or distance of the assessed points of the surface);
- for the filtration (type and nesting index of the filter, e.g. morphological filter with value of tip radius or Gauss filter with cut-off value).

If necessary, these conditions are to be specified. An ISO standard for the indication is in preparation.

For datums the association default will be standardized in ISO 5459. It will be changed from the minimum rock requirement according to ISO 5459-1988 and ISO TR 5460 to the minimum zone requirement (sometimes referred to as min. max. requirement or Chebyshev requirement). The minimum zone requirement enables the definition of common datums from more than two features, e.g. from three shafts as axis of the three associated coaxial cylinders with maximum distance from the extracted features minimized.

For multiple features with Ⓜ without datum or related to a datum with Ⓜ the recommendations given in 23.4 for designers and manufacturers should be followed in order to avoid misunderstandings in the future.

In general it can be assumed that the new approach gives more precise specifications but for the rest there is no dramatic change to the former ISO standards (ISO 1101, ISO 5459, ISO 5458, ISO 1660 and ISO 8015).

24
Synopsis of ISO Standards

The standards on Geometrical Product Specifications (GPS) have been developed by different ISO Technical Committees and only when there was a strong demand for them. Therefore in the field of GPS some standards are still missing or incomplete.

ISO has now combined the former ISO Committees into a new ISO Committee on GPS and verification (ISO TC 213).

The tasks of ISO TC 213 are to identify gaps and, if existing, contradictions in the standards and to develop future harmonized standards.

Typical tasks of the group are standards on

- definition of actual size (e.g. of cylindrical features, conical features, features composed of two parallel planar surfaces);
- definition of actual distances (e.g. of actual axes, stepped surfaces);
- definition of actual radius;
- definition of actual derived features (actual axes, actual median faces);
- definition of filter requirements in order to distinguish between roughness, waviness, form, orientation and location, size and distance;
- identification of calibration requirements;
- requirements for measuring strategy and assessment of measurement uncertainties.

The ISO Technical Report ISO TR 14 638 GPS Masterplan explains the concept of GPS and provides a Masterplan of GPS including a list of the existing standards and the standards to be issued in this field.

The concept of GPS:

- covers several types of standard, some are dealing with the fundamental rules of specification (*Fundamental GPS standards*) (e.g. ISO 8015, Fundamental tolerancing principle), some are dealing with global principles and definitions (*Global GPS standards*) (e.g. ISO 1, Reference temperature), some are dealing directly with the geometrical characteristics (*General and Complementary GPS standards*);
- covers several kinds of geometrical characteristics such as size, distance, angle, edge, form, orientation, location, roughness and waviness (*General GPS standards*) (e.g. ISO 1101, Geometrical tolerances);
- covers workpiece characteristics (tolerance classification) as results of several kinds of manufacturing processes (e.g. ISO 8062, General tolerances for castings) and characteristics of specific machine elements (e.g. ISO 965, Tolerances for threads) (*Complementary GPS standards*);
- occurs at several steps in the development of a product: design, manufacturing, metrology, quality assurance, etc.

The standards dealing with a certain geometrical characteristic form a chain of standards. The chain links deal with:

- drawing indications (symbolizations);
- definition of tolerances (e.g. tolerance zones);
- definition of actual feature characteristics or parameters (deviations);
- assessment of the workpiece deviations, and comparison with tolerance limits;
- measurement equipment requirements;
- calibration requirements.

In the future every GPS standard will contain a matrix as shown in Fig. 24.1 in order to illustrate the scope of the standard.

ISO TR 14 638 (containing the list of GPS standards) will be amended continuously. The latest issue may be purchased from the national standardization body, e.g. ANSI in the USA, AFNOR in France, BSI in the UK, DIN in Germany and DSA in Denmark (secretary of ISO TC 213).

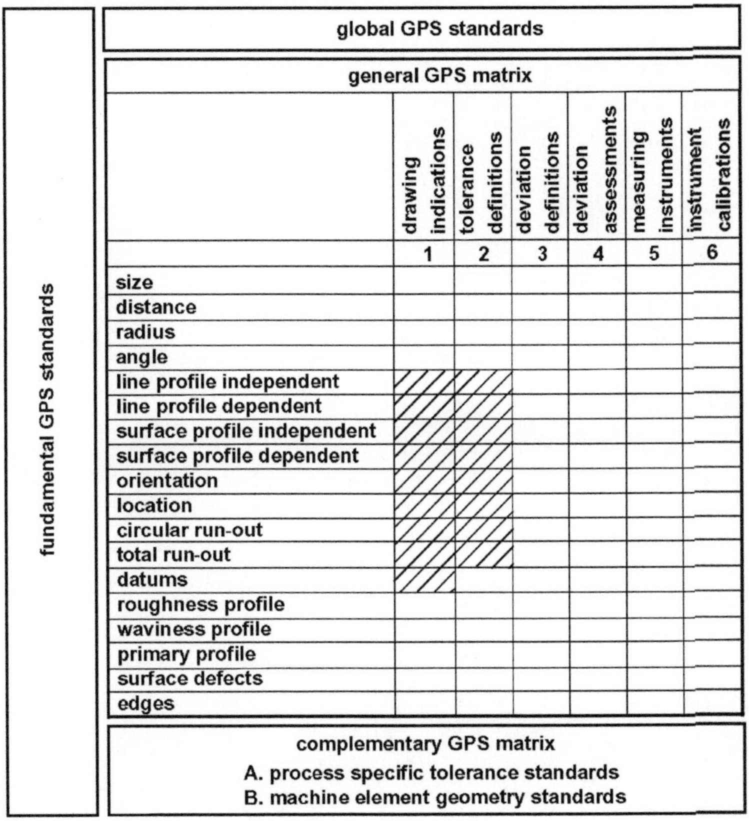

global GPS standards						
general GPS matrix						
	drawing indications	tolerance definitions	deviation definitions	deviation assessments	measuring instruments	instrument calibrations
	1	2	3	4	5	6
size						
distance						
radius						
angle						
line profile independent	▨	▨				
line profile dependent	▨					
surface profile independent	▨					
surface profile dependent	▨					
orientation	▨					
location	▨					
circular run-out	▨	▨				
total run-out	▨	▨				
datums	▨					
roughness profile						
waviness profile						
primary profile						
surface defects						
edges						
complementary GPS matrix **A. process specific tolerance standards** **B. machine element geometry standards**						

(Left vertical label: fundamental GPS standards)

Fig. 24.1 GPS matrix for the example ISO 1101

This standardization work has now become urgent, with current efforts to eliminate the trade barriers and with the growth of worldwide industrial cooperation. It is necessary, for example, to give unambiguous definitions for the programming of coordinate measuring machines and unambiguous rules for the traceability of measurements according to ISO 9001.

The British Standard BS 8888 gives an overview in the field of GPS.

Standards

ISO 286-1	ISO System of limits and fits: bases of tolerances, deviations and fits
ISO 1101	Geometrical tolerancing
ISO 1660	Dimensioning and tolerancing of profiles
ISO 2768-2	General geometrical tolerances
ISO 2692	Maximum material requirement, least material requirement, reciprocity requirement
ISO 3040	Dimensioning and tolerancing of cones
ISO 4291	Methods for the assessment of departures from roundness: measurement of variations in radius
ISO 4292	Methods for the assessment of departures from roundness: measurement by two- and three-point methods
ISO 5458	Geometrical tolerancing: positional tolerancing
ISO 5459	Datums and datum systems for geometrical tolerancing
ISO TR 5460	Geometrical tolerancing: tolerancing of form, orientation, location and run-out; verification principles and methods; guidelines
ISO 6318	Measurement of roundness: terms, definitions and parameters of roundness
ISO 7083	Symbols for geometrical tolerancing: proportions and dimensions
ISO 8015	Fundamental tolerancing principle
ISO 10 360-1	Coordinate metrology Part 1 definitions and applications of the fundamental geometric principles
ISO 10 578	Projected tolerance zone
ISO 10 579	Dimensioning and tolerancing: non-rigid parts
ISO 12 181	Measurement of roundness deviations
ISO 12 180	Measurement of cylindricity deviations
ISO 12 780	Measurement of straightness deviations
ISO 12 781	Measurement of flatness deviations
ISO 13 715	Edges of undefined shape
ISO 14 253-1	GPS, decision rules for proving conformance
ISO 14 253-2	GPS, guide to the estimation of uncertainty (in preparation)
ISO TR 14 638	GPS Masterplan
ISO 14 660-1	GPS, geometrical features; general terms and definitions
ISO 14 660-2	Extracted median line, median surface, local size
ISO 15 530-1	CMM, determining measurement uncertainty, overview
ISO 15 530-2	CMM, determining measurement uncertainty, use of multiple measurement strategies

ISO 15 530-3	CMM, determining measurement uncertainty, use of calibrated workpieces
ISO 15 530-4	CMM, determining measurement uncertainty, use of computer simulation
ISO 15 530-5	CMM, determining measurement uncertainty, use of expert judgement
ISO TR 16 570	Linear and angular dimensioning and tolerancing: +/- limit specifications – step dimensions, distances, angular sizes and radii
ISO TS 17 450-1	Model for GPS, features, characteristics, operation, specification, verification
ISO TS 17 450-2	Operators and uncertainties
VIM	International Vocabulary of Basic and General Terms in Metrology
GUM	Guide to the expression of uncertainty in measurement
ASME Y14.5M - 1994	Dimensioning and tolerancing
ANSI B89.3.1 - 1972	Measurement of out-of-roundness
BS 308 P.3	Engineering drawing practice, geometrical tolerancing
BS 3730	Assessment of departures from roundness
BS 7172	British Standard guide to assessment of position, size and departure from nominal form of geometric features
BS 8888	Technical product specification (TPS): specification
DIN 4760	Form deviation, waviness, surface roughness; system of order, terms and definitions
DIN 6784	Edges of workpieces; terms, drawing indications
DIN 7167	Relationship between dimensional tolerances and form and parallelism tolerances; envelope requirement without indication
DIN 7184	Geometrical tolerances; definitions and drawing indications (superseded by DIN ISO 1101)
DIN 7186 T.1	Statistical tolerancing; definitions, applications, drawing indications
DIN 8570 T.3	General geometrical tolerances for welded parts
DIN 40680 T.2	General form tolerances for ceramic parts in electrical application
DIN 32 880-1	Coordinate measuring technique; geometrical basics and terms
VDI/VDE 2601 T.1	Requirements on the surface structure to cover function capability of surfaces manufactured by cutting; list of parameters
TGL 39 092	Methods of measuring geometrical deviations: general principles
TGL 39 093	Methods of measuring deviations from straightness
TGL 39 094	Methods of measuring deviations from flatness
TGL 39 095	Methods of measuring deviations from parallelism
TGL 39 096	Methods of measuring deviations from roundness
TGL 39 097	Methods of measuring deviations from cylindricity
TGL 39 098	Methods of measuring deviations of the longitudinal section profile
TGL 43 041	Methods of measuring straightness deviations of axes
TGL 43 042	Methods of measuring deviations from coaxiality

TGL 43 043	Methods of measuring the radial run-out deviations
TGL 43 044	Methods of measuring the axial run-out deviations
TGL 43 045	Methods of measuring the run-out deviations in a given direction
TGL 43 529	Methods of measuring the radial total run-out deviations
TGL 43 530	Methods of measuring the axial total run-out deviations
ST RGW 301-76	Geometrical tolerances: fundamental terms
ST RGW 368-76	Geometrical tolerances: drawing indications

Publications

[1] Bäninger, E.   Das Maximum-Material-Prinzip. *VSM/SNV Norm. Bull.* **10** (1965) 109–120.

[2] Kirschling, G.   *Quality Assurance and Tolerances* (New York: Springer-Verlag 1991).

[3] Smirnow, N.W.   *Mathematische Statistik in der Technik* (Berlin: VEB Deutscher Verlag der Wissenschaften 1969).

[4] Trumpold, H., Beck, C. and Riedel, T.   *Tolerierung von Maßen und Maßketten im Austauschbau* (Berlin: VEB-Verlag Technik 1984).

[5] Wirtz, A.   Die theoretischen Grundlagen der Dreikoordinatenmeßtechnik. *Technische Rundschau* **38** (1985) 12–23.

[6] Trumpold, H., Beck, C. and Richter, T.   *Toleranzsysteme und Toleranzdesign; Qualität im Austauschbau* (München, Wien: Carl Hanser Verlag 1997).

[7] Warnecke, H.J. and Dutschke, W.   *Fertigungsmeßtechnik* (Berlin, Heidelberg, New York, Tokyo: Springer Verlag 1984).

[8] Lotze, W.   Form- und Lageabweichungen definitionsgemäß mit Koordinatenmeßgeräten prüfen. *Technische Rundschau* **31** (1990) 36–45.

[9] Neumann, H.J.   *Industrial Coordinate Metrology* (Landsberg/Lech: Verlag Moderne Industrie 2000).

[10] Barth, U.   Kombinierte Zweipunkt-Dreipunkt-Messung zum gleichzeitigen Bestimmen von Maß- und Kreisformabweichungen. *Feingerätetechnik* **33** (1984) 21–23.

[11] Wirtz, A. and Maduda, M.   Fertigungsgerechte Rundheitsmessung. *Fertigung* **5** (1972) 127–137.

[12] Heldt, E.   Komplexe Bestimmungsgröße Zylinderform und der Zusammenhang mit ihren Komponenten. *Feingerätetechnik* **33** (1984) 10–14.

[13] Trumpold, H. and Richter, G.   Unterschiede zwischen DDR-Standards und DIN-Normen auf dem Gebiet der Form- und Lageabweichungen *Feingerätetechnik* **39** (1990) 451–453.

[14] Foster: *Geo-Metrics* (IIIm, Reading Addison Wesley Publishing Company 1994).

Index